# Image Pattern
# Recognition

# Image Pattern Recognition

## Fundamentals and Applications

L. Koteswara Rao

Md. Zia Ur Rahman

P. Rohini

CRC Press
Taylor & Francis Group
Boca Raton  London  New York

CRC Press is an imprint of the
Taylor & Francis Group, an **informa** business

First edition published 2022
by CRC Press
6000 Broken Sound Parkway NW, Suite 300, Boca Raton, FL 33487-2742

and by CRC Press
2 Park Square, Milton Park, Abingdon, Oxon, OX14 4RN

ISBN: 978-0-367-64216-7 (hbk)
ISBN: 978-0-367-64224-2 (pbk)
ISBN: 978-1-003-12351-4 (ebk)

DOI: 10.1201/9781003123514

Typeset in Times
by SPi Technologies India Pvt Ltd (Straive)

# Dedication

*All Teachers & Students*

# Contents

# Preface

Recent advances in the field of digital electronics and Internet paved the way for the creation, storage and sharing of huge volumes of imagery data. An efficient image data management system is needed to retrieve and utilize the images in an effective manner. Conventional retrieval systems become inefficient with the large size of the database. Thus, there exists a dire need for an expert system that can automatically search the desired images from large repositories.

Content-based image retrieval addresses the limitations of traditional methods such as perceptional subjectivity and the amount of labor involved in the process of annotation. The primitive step in content-based image retrieval is the feature extraction, whose effectiveness relies upon the method adapted for extracting the features from a given image. It utilizes the visual contents of an image such as color, texture, shape, spatial layout, etc., to represent and index the images.

Texture analysis is used in computer vision and pattern recognition applications due to its potential in representing primary features. Many techniques are proposed to extract the texture features from an image. Since texture information always varies from image to image, there is a potential scope to identify and improve effectiveness of feature vectors in retrieval systems.

Derived from the definition of texture from standpoint of neighborhood, local patterns gained wide popularity in extracting the texture information efficiently with low computational complexity. A feature descriptor, local binary patterns (LBP) has paved the way for the creation of few simple features. Motivated by the concept of LBP and related approaches, an attempt is made in this book to introduce various methods of image retrieval using extrema patterns. Two methods: (i) Combination of color and directional local extrema patterns and (ii) combination of Gabor features and directional local extrema patterns are introduced. Local quantized extrema patterns (LQEP) are proposed to extract the texture information. The directional extrema is specifically collected from quantized geometric structures in horizontal, vertical, diagonal and anti-diagonal directions.

A modified model of LQEP is proposed in the form of local color oppugnant quantized extrema patterns (LCOQEP). LQEP information is extracted from two oppugnant planes of two different color models.

Extraction of textural information from mesh structure with the help of local mesh quantized extrema patterns is another proposal discussed in this book. Evaluation of the above models is carried using benchmark datasets. The performance is evaluated in terms of average retrieval precision and average retrieval recall. Satisfactory improvement is observed and also reported in this book.

This book is useful for B. Tech and M. Tech students of image and signal processing of any university.

# Acknowledgments

Our sincere thanks are due to Dr. L Pratap Reddy, Professor of ECE Department, JNTUH, for his practical analysis and thorough discussions during the preparation of conceptual works.

We wish to thank all our research associates who have participated in many discussions and for sparing their valuable time and effort in reviewing this book for technical perfection.

We also appreciate the good company of the people at the Department of Electronics & Communication Engineering and the Management of Koneru Lakshmaiah Education Foundation (K. L. University), Hyderabad and Vijayawada campuses for their continuous support to complete this work.

# Authors

**Dr. L. Koteswara Rao** received his Ph.D. from JNTUH, Hyderabad; M.E. from College of Engineering Andhra University, Visakhapatnam and B. Tech from Jawaharlal Nehru Technological University, Kakinada, India. He has 18 years of teaching experience. He published more than 30 papers in various reputed National, International Journals and Conferences. He has guided 25 M. Tech projects and about 60 B. Tech projects in different fields. His research interests include image processing, signal processing, embedded systems and IoT.

**Md. Zia Ur Rahman** received the M. Tech and Ph.D. degrees from Andhra University, India. He has been with the Department of Electronics and Communication Engineering, Koneru Lakshmaiah Educational Foundation, K. L. University, Guntur, India, since 2013 as a professor. He has authored or co-authored more than 120 articles on indexed international conferences and journals, as well as five international books. His main areas of research are artificial intelligence and blockchain for health care systems, cognitive communications, machine learning, measurement technology and signal processing applications. He serves an associate editor for Measurement Journal (Elsevier, NL), Measurement: Sensors (Elsevier, NL), Measurement: Food, (Elsevier, NL), IEEE Access (USA) and editor in chief for International Journal of Electronics, Communications, and Measurement Engineering, USA. He is a senior member of IEEE, USA and a fellow member of Institute of Engineers (India). He is a scientific consultant for several national and international institutions.

**P. Rohini** received her B. Tech and M. Tech in computer science and engineering from JNTU. She has 14 years of teaching experience. Currently, she is working as an assistant professor in the Department of CSE, Faculty of Science and Technology, IFHE, Hyderabad. Her research interests include image processing, data mining and deep learning.

# 1 Introduction

## 1.1 DATA RETRIEVAL

Internet has become a source for massive amount of data being generated every moment. The data includes the text or words, images, videos, slides, models, workflows, artifacts and many more. Prominence of imagery data is increasing everyday due to the advances in multimedia technologies and the Internet. Manual processing of data is a tedious task, and hence there is a need for an image management system to organize the same.

In the early years, image retrieval used to involve manual elucidation and a database management system to store the information. Metadata associated with every image has a key role in indexing and retrieval. However, the method of manual elucidation faces two problems, primarily when the volume of image collections is very much high. The first difficulty is the huge quantum of labor involved in the annotation process of images manually, while the other one is the variation in the perception of inherent image content. Hence, there is a possibility of getting many mismatches in the retrieved results if these two problems are not addressed thoroughly. The process of manual image elucidation became much more complex during the early '90s due to more image collections owing to the technological inventions in digital image sensor field.

## 1.2 CONTENT-BASED IMAGE RETRIEVAL SYSTEM

To surmount the issues in traditional retrieval methods, an automated approach named content-based image retrieval was originated. Unlike the manual annotation method, indexing and subsequent retrieval is done based on the inherent content such as texture and color of the image.

In content-based image retrieval framework, description of visual data is done by generating feature vectors of various dimensions and storing the same in a database. To get the image retrieved, the user provides the imagery data in query form. Then, the system creates equivalent delineation of query image. Distances between query and database vectors are calculated so as to determine the similarity levels. In the next step, an indexing scheme is followed to search the entire database. Recently, retrieval systems are designed by keeping the user in the loop to specify the relevance in order to enhance the process and achieve semantically effective outcome. Figure 1.1 depicts a generic structure of content-based image retrieval system.

Generally, feature descriptors used in content-based image retrieval are either local or global. The whole image is considered to generate a global feature descriptor, whereas objects or regions are represented by local feature descriptor. These are further categorized into spatial and transform domain features. Ever since the

DOI: 10.1201/9781003123514-1

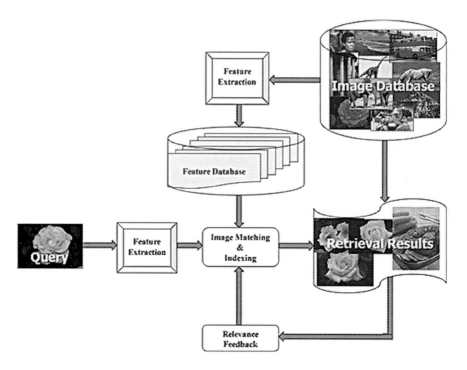

**FIGURE 1.1**    Architecture of content-based image retrieval.

introduction of this method, researchers across the world devised several approaches for various applications. The researchers studied the area of image retrieval in two different perspectives, mainly based on metadata and visual data.

An extensive review of various image retrieval systems, such as AltaVista Photo finder, Advanced Multimedia Oriented Retrieval Engine (AMORE), ASSERT, Berkeley Digital Library Project (BDLP), CANDID, CBVQ, CHROMA, FIDS, Image Rover, Multimedia Analysis and Retrieval System (MARS), Query By Image Content (QBIC), etc., can be found in Veltkamp and Tanase (2002) and Kherfi et al. (2004).The philosophy of CBIR is available in Smeulders et al. (2000).

### 1.2.1  IMAGE DATABASES

To measure the effectiveness of our methods as compared to other methods, the following natural and texture image databases are considered.

* Natural Image Database

Two image repositories, (i) Corel database and (ii) ImageNet, are considered to evaluate proposed methods against relevant methods. Corel database has imagery data with varying contents and dimensions. Hence, researchers believe that Corel repository fulfills majority of requirements to assess any research work. Corel-1000 contains 1000 images of size 384 × 256 or 256 × 384. The images are arranged into

**FIGURE 1.2** Image samples from Corel-1000 repository.

10 categories, each with 100 images. Figure 1.2 shows the image samples of Corel-1000 natural image database.

Corel-5000 is a subset of Corel database. It contains 5000 natural images gathered from 50 different domains. Corel-10000 database contains 10,000 natural images collected from 100 varieties of domains. Each domain possesses 100 images of size $384 \times 256$ or $256 \times 384$.

The ImageNet-25k database is a composition of 25,000 various natural images divided into 25 different categories. Each category is a collection of 1000 images. Figure 1.3 depicts few images from ImageNet-25k image repository.

- Texture image repository

The MIT Vision Texture (VisTex) database contains 40 colored texture images of size $512 \times 512$. Further, each one is partitioned into 16 sub-images of dimension $128 \times 128$ without overlap, thereby producing 640 ($16 \times 40$) texture image database. Image samples of MIT VisTex database are provided in Figure 1.4.

## 1.2.2 EXTRACTION OF FEATURES AND THE CREATION OF FEATURE DATABASE

This is a prominent stage of image retrieval framework. Success of retrieval heavily relies on how close these features represent an image. A perceptible content, such as shape, texture and color, forms a base for feature vector. Image features are broadly

**FIGURE 1.3** Image samples from ImageNet-25K repository.

**FIGURE 1.4**    Samples of MIT Vision Texture database.

categorized into high and low levels. Features like color, shape and texture fall under the first category while the keywords, text descriptors, etc., come under second category. In this book, designs of the following low-level features-based frameworks are primarily focused.

- Improved Directional Local Extrema Patterns
- Local Quantized Extrema Patterns
- Local Color Oppugnant Quantized Extrema Patterns
- Local Mesh Quantized Extrema Patterns

Subsequent to the feature extraction, a database of features is created for the retrieval. In fact, feature database creation is an offline process in any content-based image retrieval model.

### 1.2.3    QUERY IMAGE

User can make a query in one of the following ways in image retrieval system (R. Dutta et al. 2008).

- **Keywords**: User provides keywords to search a particular image/object, i.e. flower, animal, etc. The search engines like Yahoo and Google use this method for searching the images.
- **Images**: User provides an example image to search for a similar image. Recently, Google search engine updated the system by incorporating this scheme.
- **Graphics**: In this search format, user provides a drawing or sketch of the image for which a search has to be made.
- **Composite**: By combining any one of the aforesaid schemes, an interactive querying is done as used in relevance feedback systems.

In this book, query is presented using an example image.

### 1.2.4    IMAGE MATCHING AND INDEXING

In the first step, user provides an example image for which a similarity search has to be found. The system then converts the query image into its internal format to create a vector of features. Feature vectors are compared to determine the similitude. Upon calculating the distances between the vectors, ranking method is followed to find out the similarity. In this manner, retrieval of images is performed. Modern retrieval systems keep users in the loop to refine the search, which is termed as relevance feedback.

Similarity and performance measures used in the current research work are specified in the next section.

### 1.2.5  Similarity Distance Measures

Similarity measure is another vital step in content-based image retrieval. Different types of similarity measures such as Euclidian distance, Histogram intersection, Mahalanobis distance are applied to get similar images. A distance function or metric 'S' by definition is non-negative, symmetric and satisfies the triangular inequality. It has the property that S(K,L)=0 if and only if K=L Dohnal et al. (2003). Not only the selection of appropriate feature vectors but also the selection of suitable distance metric impacts the performance of content-based image retrieval system. Following similarity measures are used to evaluate algorithms.

- City-block or Manhattan or L1 distance

In the Manhattan or L1 or city block distance, the absolute differences of the features are considered, and hence it is inexpensive than the Euclidian distance method.

Mathematically, it is expressed as
*Manhattan distance metric*

$$Dist\left(r,I_1\right) = \sum_{k=1}^{Lg} \left|v_{DB_{jk}} - v_{q,k}\right| \tag{1.1}$$

$L_2$ Distance or *Euclidian* distance
*Euclidean distance metric*

$$Dist\left(q,b\right) = \left(\sum_{m=1}^{Ln}\left(v_{b_{jm}} - v_{q,m}\right)^2\right)^{1/2} \tag{1.2}$$

It has to be noted that the computation of square root is the most expensive operation.

- $D_1$ Distance
  $d_1$ distance *metric*

$$Dist\left(q,b\right) = \sum_{m=1}^{Ln}\left|\frac{v_{b_{jm}} - v_{q,m}}{1 + v_{b_{jm}} + v_{q,m}}\right| \tag{1.3}$$

- Canberra distance

*Canberra* distance *measure:*

$$Dist\left(q,b\right) = \sum_{m=1}^{Ln} \frac{\left|v_{b_{jm}} - v_{q,m}\right|}{\left|v_{b_{jm}}\right| + \left|v_{q,m}\right|} \qquad (1.4)$$

where $v_{b_{jm}}$ is $m^{th}$ attribute of $j^{th}$ image of the repository $|b|$ and $v_{q,m}$ represents the $m^{th}$ feature of query image $q$.

## 1.2.6   RELEVANCE FEEDBACK

Inputs from the user are also incorporated in a supervised learning method termed relevance feedback. This approach improves the retrieval relevance. Upon completing the search, the system provides retrieval results. Then, user identifies the relevant and non-relevant examples of query image. User's feedback enables refinement of results to produce new output. The main aspect of relevance feedback system is the inclusion of negative and positive examples for the refinement of searched images so as to adjust for similarity (Su et al. 2003).

## 1.2.7   PERFORMANCE MEASURES

A crucial issue in content-based image retrieval is evaluation of retrieval performance (Muller et al. 2001). Researchers used various methods to evaluate the performance of systems. Performance metrics such as precision and recall are implemented for information retrieval and in the content-based image retrieval too. However, it is worth noting that no common method exists in image retrieval (V. N. Gudivada and Raghavan 1995).

In this book, evaluation metrics such as average retrieval precision (ARP) and average retrieval recall (ARR) are applied to estimate the performance of each retrieval system. Introduction of the evaluation measures is available in (Huijsmans and Sebe 2005).

Precision of image $I_q$ is expressed as

$$PR\left(Iq\right) = \frac{Relevant\ Images\ retrieved}{Total\ No.of\ Images\ Retrieved} \times 100 \qquad (1.5)$$

$$Group\ Precision\left(GrPR\right) = \frac{1}{X_1}\sum_{a=1}^{X_1} PR\left(I_q\right) \qquad (1.6)$$

$$Avg.Retrieval\ Precision\left(ARP\right) = \frac{1}{Y_1}\sum_{b=1}^{Y_1} GrPR \qquad (1.7)$$

Recall is expressed as

$$R\left(I_q\right) = \frac{No.of\ relevant\ images\ retrieved}{Total\ no.\ of\ relevant\ images} \tag{1.8}$$

$$Group\ Recall\left(GR\right) = \frac{1}{X_1} \sum_{a=1}^{X_1} R\left(I_q\right) \tag{1.9}$$

$$Avg.Ret.rate\left(ARR\right) = \frac{1}{Y_1} \sum_{b=1}^{Y_1} GR \tag{1.10}$$

Here, $X_1$ and $Y_1$ represent no. of pertinent images and no. of categories, respectively.

## 1.3   ORGANIZATION OF THE BOOK

The rest of the book is organized into eight chapters.

Specific review on various image features that are prominently identified by various researchers is presented in Chapter 2. Information pertaining to individual features like texture, shape and color and extraction methods is listed here. Impact of multiple features on retrieval systems and the limitations of these approaches are discussed. The importance of local feature extraction methods from the standpoint of illumination, rotation and scaling factors is shown in this chapter. The problem definition in terms of the need for feature exploration of local features is discussed followed by the methodology adopted in the present work.

Chapter 3 introduces two models of directional local extrema patterns, viz., (i) integration of color and directional local extrema patterns and (ii) integration of Gabor features and directional local extrema patterns. Improvements of these models from the base model of directional patterns with the induction of significant information in the form of extrema with specific reference to color and Gabor features are presented. Evaluation in the form of ARP and ARR is carried out on Corel-1k repository.

In Chapter 4, local quantized extrema patterns are introduced. Framework to extract the features using quantization method is presented. ARP and average recall values from Corel-1k, Corel-5k and MIT VisTex database are listed. Evaluation results in terms of improvement in precision and recall parameters are discussed.

Chapter 5 introduces local color oppugnant quantized extrema patterns. Local quantized extrema collected from two oppugnant color planes are used in this method. Experimental results from Corel-1k, Corel-5k,Corel-10k and ImageNet-25k are provided.

Chapter 6 describes local mesh quantized extrema patterns. A mesh structure is created by considering the pixels at alternate positions. Evaluation is carried out on MIT Vis Tex and Corel-1k databases.

Chapter 7 provides few more feature descriptors for image retrieval. Most of the methods are tested on Corel database.

Chapter 8 concludes the entire work and the detailed summary. The tradeoff between the dimensionality of feature vector and the retrieval accuracy in the form of precision and recall is specified in the chapter. Future directions are presented in continuation.

# 2 Features Used for Image Retrieval Systems

## 2.1 INTRODUCTION

Texture, color and shape are the primitive features in content-based image retrieval systems. A detailed review of integration of different low-level features, i.e. texture, shape and color, is presented in this chapter.

## 2.2 COLOR FEATURES

With the advancement of technology, substantial changes in lifestyle have taken place. In the context of present work, a noteworthy example is watching color movies, where the degree of enjoyment one can perceive is higher than viewing black and white movies. Color gives more information about a scene in addition to the beauty. Same information is considered as a tool to retrieve images. Many image search schemes were proposed based on color feature.

Swain and Ballard (1991) presented a method of color histogram in 1991 and formulated histogram intersection distance measure to estimate distances among histograms. A holistic approach like Global Color Histogram (GCH) was applied in image retrieval to obtain the frequency of existence of each color in image. Even though, it followed translation, rotation and scale invariance properties, it couldn't provide the spatial information due to which a false retrieval may take place.

Stricker and Oreng (1995) introduced color indexing methods. In one of the approaches, cumulative color histogram was applied to get spatial information. In other approach, only three moments pertaining to each color space were extracted instead of complete color distributions. This method based on color moments avoids the impact of quantization but lacks spatial information.

In this context, Idris and Panchanathan (1997) introduced a method called vector quantization to compress the image, create a code word and obtain the histogram of code words. The histogram obtained at the end becomes the feature vector to match images. On the same lines, Lu and Burkhardt (2005) introduced a method to retrieve color images by adapting the combination of Discrete Cosine Transform (DCT) and vector quantization technique.

In Qiu (2003), explained compression of image by Block Truncation Coding (BTC) method. Further, Lu and Teng (1999) suggested the use of compressed image data for image retrieval to capture the spatial relationship among the pixels while indexing the image. Another way to collect the color feature by using a variant of $k$-means algorithm was introduced by Su and Chou (2001).

Global histogram, which is used to represent the color information, does not provide any spatial structures. To overcome the limitation, Huang et al. (1997)

DOI: 10.1201/9781003123514-2

introduced color correlogram method that yields three-dimensional matrices to depict the probability of identifying the pixels related to any two given colors. However, computationally it is unworthy for retrieval process.

Pass et al. (1997) introduced color coherence vectors, in which histograms were built by incorporating the spatial information. In this method, image was initially smoothed to eliminate small differences in pixel values. The pixels within the histogram were separated into two parts: a coherent vector and a non-coherent vector. When a pixel lies in the large class of connected pixels of a color, it becomes coherent pixel.

Rao et al. (1999) proposed a modification to color histogram by creating annular, angular and hybrid color histograms. This method of histograms outperformed the traditional color histogram and color coherent vector.

Chang et al. (2005) introduced an approach having lower dimension of feature vector when compared to color correlogram. It considered the variation in color due to change of illumination and camera's view angle. Another method for extracting the spatial information was suggested in Chen et al. (2004), where image was first partitioned using JSEG technique to capture the color properties of image. Intuitive color of the segmented area was subsequently transformed into a different form, i.e. semantic color name. This method surpassed the conventional histogram approach. Problem of segmentation was rectified in Wang et al. (2009) using salient point method. Salient points were selected using multi-scale Harris-Laplace detector, and around these points, color histogram of local feature regions was constructed to serve as a feature vector in the retrieval process.

Park et al. (2000) introduced a modified local Edge Histogram Descriptor (EHD) for MPEG-7 by constructing the semi-local and global edge histograms directly from local histogram bins. Cinque et al. (1999) introduced the spatial chromatic histogram which provides the spatial information of pixels having the same color. Li (2003) added spatial information by running sub-blocks with different similitude weights, and color features were extracted for every sub-block. In the next step, the resemblance between corresponding sub-blocks was calculated.

Yoo et al. (2005) used two sets of features to filter out the irrelevant images, i.e. Major Colors' Set signature for color data and Distribution Block Signature for spatial inputs. To extend this method, filtered out images were compared with query image using global color set (GCS) and a new feedback technique based on quad modeling (QM) was proposed.

Jiang et al. (2006) proposed an efficient dominant color extraction algorithm directly in DCT compressed domain. It is the fast processing method for content description of compressed images and videos. Problem of fixed bin in global histogram method was solved by Ma et al. (2010) using a new variable bin size distance for histogram features.

Lin et al. (2011) conceptualized a new scheme for image retrieval with spatial information by extracting color spatial distribution (CSD) features in three ways. Initially, all images were quantized using $k$-means algorithm and three CSD features were extracted from the quantized image. A genetic algorithm was used to attune the parameters of CSD features. At query time, filtering was attained by using cluster-based technique.

Conci and Castro (2002) studied the impact of different distance metrics (city-block, Euclidean, histogram intersection, average color distance and the quadratic distance form) on image mining depending on color feature.

## 2.3 TEXTURE FEATURES

One among the important features considered in image retrieval is texture. Using texture feature, we can discriminate the images of leopards and tigers because the leopard has black spots and the tiger has strips; in other words different texture patterns exist.

Due to the variations in intensity and color, certain repeated motifs called visual textures are formed as specified in Tuceryan and Jain (1993). Four approaches are used in analyzing the textures of a region. They are statistical, structural, signal processing and stochastic texture modeling methods. Spatial arrangement of intensities and few statistics of the distribution are used in statistical methods. Local feature exploration methods come under statistical methods (Srinivasan & Shobha 2008; Haralick 1979).

Statistical methods are further categorized into statistics of first order (single element), second order (two elements) and higher order (above two picture elements) depending upon spatial alignment of pels. Most commonly used statistical methods are Haralick's Gray-Level Co-occurrence Matrix (Haralick et al. 1973a), gray-level differences (Weszka et al. 1976) and auto-correlation feature (Tuceryan & Jain 1998; Ojala et al. 2001). Structural methods deal with some primitives and certain placement guidelines. Some of the structural methods are blob method (Haralick et al. 1973b), Voronoi polygons, etc. (Tuceryan & Jain Chatterji 1990).

Local binary patterns approach (Ojala et al. 2002) is a unified technique that incorporates the statistics of local structures reflecting the combination of structural and statistical methods. An effective method to delineate the texture is modeling, in which the framework is built using few specifications to generate the type of texture of interest. Fractal analysis (Pentland 1984) and Markov Random Field (MRF) model (Ojala et al. 1996) were used for this purpose. Frequency content of image was considered in signal processing methods to get frequency detail of an image. The most commonly used signal processing methods are Fourier Transform, Gabor Transform and Wavelet Transform. Relevant literature on texture features is continued below.

Tamura et al. (1978) instituted six properties of texture such as contrast, coarseness, line likeness, directionality, roughness and regularity popularly called Tamura features. Haralick et al. (1973a) proposed Gray-Level Co-occurrence Matrices. This method was elevated as the most popular, extensively used texture feature analysis method. Similarly, various structural and statistical texture analysis methods were discussed in Haralick (1979).

Based on the technique mentioned in Moghaddam et al. (2003), in Saadatmand and Moghaddam (2005), proposed an approach named Wavelet Correlogram (WC) in which they applied three-scale disintegration algorithms on every image and subsequently implemented four-level quantization on various sub-bands of wavelet transform coefficients. Quantized sub-bands were made to undergo auto-correlogram operation resulting into feature vectors of the corresponding image.

Saadatmand and Moghaddam (2005) improved Quantization thresholds of sub-bands by employing a genetic algorithm. The means of feature extraction remains unchanged as in the WC. Feature vector length of a given image was 96 in this case. Moghaddam and Saadatmand Tarzjan (2006) obtained a noticeable improvement in the evaluation metrics compared to earlier methods by replacing the Wavelet Transform with the Gabor Transform.

Mandal et al. (1999) applied the Daubechies wavelet in three scales to get the converted output. Histograms related to the wavelet coefficients from every sub-band were utilized as the feature vectors in indexing the images.

Do and Vetterli (2002) devised Discrete Wavelet Transform (DWT)-dependent feature with Generalized Gaussian Density (GCD) and Kullback Leibler Distance. This method was experimented on VisTex database and exhibited an improvement in the performance as compared to conventional approaches such as the Wavelet frames pyramid transform.

Manjunath and Ma (1996) applied Gabor wavelets in image retrieval applications. It outperformed the traditional Pyramid Wavelet Transform technique, Multi-Resolution Simultaneous Autoregressive Model (MR-SAR) and Tree-structured Wavelet Transform methods on Brodatz texture database. Ahmadian and Mostafa (2003) applied Gabor Wavelet Transform for texture classification which exhibited superiority in performance over the dyadic Wavelets. In dyadic Wavelets, some mid-band information gets lost, while the Gabor Wavelet retains the same.

Ursani et al. (2007) compared Discrete Fourier Transform with Gabor Transform for texture information retrieval. They established that Gabor Transform outper-formed DFT approach for images containing Gaussian, Salt and Pepper noise. Challa et al. (2007) proposed modified Gabor Transform to tackle the directional sensitivity problem of the Gabor transform. Experimental simulation results demonstrated that the modified Gabor based method was effective for Content-Based Image Retrieval (CBIR) and thus showed better retrieval performance.

Han and Ma (2007) proposed the rotational and scale-invariant Gabor transform for CBIR with the improved performance as compared to the standard Gabor trans-form technique. Vo et al. (2006) proposed the complex directional filter bank (CDFB) combined with the Laplacian pyramid for the feature vector generation. Experimental results provide good overall performance in texture retrieval rate as compared to few other directional transforms such as the Gabor Wavelet, the contourlet and the steer-able pyramid.

Rallabandi and Rallabandi (2008) proposed the multistate Wavelet-based hidden Markov trees (MWHMT) for rotation-invariant texture retrieval with Kullback Leibler distance to determine the similarity of textures. Experimental work per-formed on the Brodatz texture database provided better results in the form of preci-sion and recall than other Wavelet approaches such as the Wavelet packet signature (WPS), the polar Wavelet energy signature (PWS), the rotated complex Wavelet fil-ters and the DWT Wavelet pyramids.

Rajavel (2010) proposed Directional Hartley Transform in two ways, viz., Wrapping-based Directional Hartley Transform and Overlapping-based Directional Hartley Transform. First method adapted Fourier integral operator tiling scheme and second technique adapted Ridgelets method.

Wavelet Transform finds a good place in the texture analysis but suffers from the directional problem. In texture classification, specific directional information of an image is necessary. A two-dimensional (2D) DWT splits data into four sub-bands. HH sub-band provides diagonal data. However, separation of 45° and 135° direction is hard.

By rotating the standard 2D DWF filters, Kim and Udpa (2000) introduced a new methodology called rotated Wavelet filter (RWF) to address the problem of diagonal information. Results depict that the combination of discrete Wavelet filters and RWF provides considerable classification accuracy on the Brodatz repository.

Kokare et al. (2007) applied the same concept and achieved improved performance over the DWT and the Gabor transform. Kokare et al. (2004) proposed less complex and faster Cosine -Modulated Wavelet Transform to extract texture. Wavelet Transform is widely used in image processing. However, it has the limitation in efficiently representing any object having anisotropic element such as lines or curvilinear structures.

The Curvelet transform preserves the edge information accurately as compared to Wavelets and Gabor features. Related material and the applications of the Curvelet transform can be found in Starck et al. (2002), Donoho and Duncan (2000), Fadili and Starck (2007), Sumana et al. (2008), Kinage and Bhirud (2010), Mandal et al. (2009) and Mandal et al. (2007).

Rivaz and Kingsbury (1999) conceptualized Dual-Tree Complex Wavelet Transform associated with the approximate shift invariant and six directional information ±15°, ±45°, ±75° properties. Even though, the Gabor transform finds its place in many applications of image texture retrieval, it has the disadvantage of slow computation. Complex Wavelets, a slightly modified form of the Wavelets, are approximately shift invariant when compared to the DWT.

Rivaz and Kingsbury (1999) suggested the usage of Complex Wavelet Transform in texture extraction. Experimentation showed that Complex Wavelet Transform achieves a similar performance to the Gabor Wavelet Transform with higher speed and superiority in performance as compared to a normal Wavelet Transform.

Ahmad Fauzi and Lewis (2008) used the multi-scale sub-image comparing technique combined with Discrete Wavelet structures to retrieve texture images without making any image partitions. In the work by Guo and Hatzinakos (2007), integration of DWT and Radon Transform was used to propose an image hashing method. Two different images possess two different hashes, and hence this method is suitable for retrieval.

Zuo and Cui (2009) proposed fractional Fourier transform (FRFT)-based image hashing technique for CBIR application. Concept of FRFT is available in Almeida (1994) and in Shinde and Gadre (2001).

## 2.4 LOCAL FEATURES

He and Wang (1990) derived a texture spectrum based on the presence of texture in local neighborhood. Subsequently, Ojala et al. (2001) devised local binary patterns to describe texture and these patterns were extended for rotational invariance (Pietikainen et al. 2000).

Pietikainen et al. (2000) introduced rotational invariant texture classification with feature dispensation. Ahonen et al. (2006) and Zhao and Pietikainen (2007) applied local binary patterns in face identification and expression analysis. Heikkila and Pietikainen (2006) applied local binary patterns in background detection and modeling. Huang et al. (2004) proposed extension of local binary patterns for shape localization. Heikkila et al. (2009) applied local binary patterns to describe regions of interest.

Integration of local binary patterns and Gabor features for texture segmentation is found in the work by Li and Staunton (2008). Zhang et al. (2010a) devised Local Derivative Patterns for facial image recognition and treated local binary patterns as non-directional, first-order patterns. Takala et al. (2005) proposed block-based texture feature that uses local binary patterns as tool to describe the content.

To describe the interest regions, modified form of popular local binary patterns named center-symmetric local binary pattern was integrated with Scale Invariant Feature Transform in Takala et al. (2005). Yao and Chen (2003) introduced two forms of histograms for Local Edge Patterns, viz., LEPSEG and LEPINV. The first method was susceptible to the changes in rotation and scale, whereas the second method was resistive to rotation and scale.

Local ternary pattern method (Tan and Triggs 2010) was proposed for face recognition at various illuminations. Subramanyam et al. (2011, 2012) devised varieties of patterns such as Local Maximum Edge Patterns, Local Tetra Patterns, Directional Local Extrema Patterns, Directional Binary Wavelet Patterns and Local Ternary Co-occurrence Patterns.

Reddy and Reddy (2014) improved Directional Local Extrema Patterns method by including magnitude data associated with intensity values of an image for image retrieval. Hussain and Triggs (2012) introduced local quantized patterns (LQP) for visual perception. In this method, quantization was followed in four major directions.

## 2.5 SHAPE FEATURES

Shape is one among primitive features in CBIR. Description of shape features usually done after partitioning the image into shapes or regions. Due to the difficulty faced in achieving perfect segmentation, usage of shape features is confined to specific utilizations when objects or regions are readily available. Shape is represented in boundary-based and region-based form. Different descriptors such as the Fourier descriptor, the statistical moments, etc., are used for the shape representation.

Mehtre et al. (1997) analyzed the effectiveness of different shape measures for content-based retrieval of images. They discussed the shape measures such as outline-dependent features, region-dependent features and combined features. Zhang and Lu (2002) conceptualized Fourier descriptor, using a 2D Fourier Transform on shape image. Zhang and Lu (2003) compared Fourier descriptors with curvature-scale space descriptors which were used as the curve-shape descriptors for image retrieval.

## 2.6 MULTIPLE FEATURES

From childhood, everyone knows that the combination of two or more good things gives better result. The same philosophy is equally applicable to image retrieval applications. In retrieval, single feature has its own limitations. For example, color features can't differentiate between the black buffalo and the black car if both images have the same size. Here, if we add the shape feature in addition to the color, definitely one can expect the correct result. The same story will repeat in case of tigers and leopards. Both animals have same color but can be differentiated on the basis of texture features because leopards have black spots but the tigers have black strips. In CBIR literature, numerous methods are available with combinations of the texture, color and shape. Some relevant methods are discussed in the following course.

Recently, methods on integration of color and texture gained attention. Jhanwar et al. (2004) conceptualized Motif Co-occurrence Matrix for image retrieval. Color Motif Co-occurrence Matrix was implemented with MCM on red (R), blue (B) and green (G), color planes separately. Fusion of color and texture was devised by Lin et al. (2009).

Vadivel et al. (2007) introduced a combined Intensity Co-occurrence Matrix and color features for retrieval application. Subsequent to the detailed analysis of HSV space, weighted functions were considered to explore contributions of color and gray levels of a pel.

Manjunath et al. (2001) introduced three descriptors separately for color and texture features. The final draft committee of MPEG-7 standard approved the same. These descriptors were diligently examined and evaluated according to MPEG-7 experiment standards to prove the effectiveness in broad range of applications.

In the similar fashion, Zhu et al. (2000) used the key-block-based method to yield a solution for image retrieval that resembles text-based information retrieval. Key-block codebooks were generated using the vector quantization methods. Objects and the features were derived on the basis of the appearance and frequency of key blocks within images.

With the same approach, Liu and Zhang (2007) proposed an effective method depending on the composition of the key blocks and the interest points to resolve the problems in feature representation, which has the comprehensive contribution to the local region and the object border in an image.

Mehtre et al. (1998) used fusion of color and shape features on clustering method for trademarks and artificial images. Similarly, spatial information with color is provided in Mehtre et al. (1998).

Prasad et al. (2004) combined color and shape features at local level by partitioning the image into nine equal blocks. Shape was represented by new descriptor. It was unvarying to rotation, translation and scaling. Color features were extracted from the dominant color clustering method. Further, they used hashing technique to store combined indices of images. Images with same hash structure form a cluster and the indices were used to compare images for similarity, thus reducing the search space and time.

A new method named SNL was presented in Sridhar (2002), where each image was partitioned to get the features including color (color histogram of segmented region), shape (height to width ratio), spatial locality and size of the region. A non-metric distance measure named Integrated Region Matching (IRM) was used to match the regions by comparing the content, the spatial position and the shape. Further, they showed that SNL outperforms the GCH and the color-based clustering (CBC) in terms of precision and recall.

Pour and Kabir (2004) proposed new retrieval scheme by using local chromatic and local directional dissemination of gradient. Initially, image was partitioned into 4x4 sized blocks and every block was classified into uniform or non-uniform, based on its intensity gradient. For Uniform block, average of each component was calculated to decide representing color to that block. Subsequently, a histogram of unicolor uniform blocks of the image was built. In case of the non-uniform blocks, the features were in the color representatives and histograms of bi-color non-uniform blocks. Similarly, for representing shape information of image, inter-block directional changes of gradient were determined and a histogram of directional variations in gradient was produced.

Qi and Han (2005) devised a new integrated approach using color, texture, global and semi-global EHD. Individual fuzzification was adapted on each of the regions in image to blend segmentation-related ambiguities into the retrieval algorithm.

Liu and Yang (2008) proposed a new approach of co-occurrence matrix, i.e. texton co-occurrence matrix (TCM), using the Julesz textons theory and the correlogram concept which yields better outcome as compared to Gray-Level Co-occurrence Matrix and color correlogram. Further, improvement in TCM was obtained by introducing the multi texton histogram (MTH) method (Liu et al. 2010). It takes the positive outcomes of the histogram of the co-occurrence matrices. Further, they created a new feature named micro-structure descriptor (MSD) (Liu et al. 2011), based on an edge orientation similarity.

Jeena Jacob et al. (2014) devised the local oppugnant color texture pattern, a modified form of Local Tetra Patterns to differentiate information from the spatial patterns of the oppugnant channels of a local area. The directional information was derived from two different pixels. The LOCTP attempted to utilize the compatibility between texture and color, thereby enabling the framework to include user's perception.

Karakasis et al. (2015) introduced a system which depends on affine moment invariants to describe the local areas in the image. A representation named bag of words was used to store the produced moments, which was finally used as a feature vector. Image moment invariants were selected in the specified work due to the ability to get compact representation. Three different methods were followed to evaluate the same. The results were better as compared to popular local descriptors, making the proposed work to set a benchmark for future image moment local descriptors.

Kauppi et al. (2015) proposed that human brain action could be involved in controlling the future information retrieval frameworks. In this method, a practicality study was conducted to know the prediction of similarity of objects based on the neural activity. They analyzed magneto encephalographic and some signals out of nine different parts to create a subset. They came up with three discoveries which created a new platform for designing an interactive system of retrieval and for using return response both from the activity of the brain and the eye movements.

Yang et al. (2015) proposed a method called the learning salient visual words for moving and the scalable image retrieval. The retrieval performance was improved by including the soft spatial verification to rearrange the results. This method reported a less data transmission and a less computational cost.

Zhou and Fan (2015) introduced an automatic image-text alignment method to obtain more efficient indexing and retrieval of a large number of web images by aligning images with their relevant text terms or phrases. In the first step, a large number of web pages were crawled and segmented into a set of image–text pairs. Based on the image similarity, sets of web image clusters were formed and the random walk was applied on phrase correlation system to get better and precise image–text alignment.

Vipparthi and Nagar (2014a) proposed new descriptor, named Directional Local Motif XoR Patterns (DLMXoRPs). Exclusive-OR operation was implemented on motif images that were missing in previous works. To focus on the benefits of the method, they made comparisons with Motif XoR pattern.

Lin et al. (2014a) introduced a genetic algorithm feature selection (GAFS) for categorization and retrieval. Results were primarily derived from three features: GHAM, ACH, and AMCOM. The GAFS was yielding a better solution at the price of increased computational complexity. It was reported that the usage of GAFS system reduces the count of features and increases the retrieval efficiency.

Kumar et al. (2014) introduced a new technique to determine the document image systemic resemblance in the applications of retrieval and classification. They first created a codebook of Speeded Up Robust Features (SURF) descriptors derived from a group of representative images. A recursive partitioning method was adapted to encode the spatial relationship between the document images. Histograms were built in each segmentation. This method outperformed the previous approaches even for a limited datasets of images.

Lin et al. (2014b) proposed a method to design a fast K-means algorithm to be used with clustering data to reduce the time spent on training the image cluster centers. It overcame the cluster center retraining issue, as more numbers of images were added to the database continuously. The gray and color image sets were used for the performance comparisons.

Megumi et al. (2014) devised a camera-dependent method for the digital recuperation of handwritings on an ordinary paper. The method was recognized by the following two outcomes: (1) there was no need of special device except a camera pen to retrieve handwritings and (2) even for handwriting on a printed document, the method was able to localize it onto an electronic equivalent data.

Based on multi-scale run length histograms, Gordo et al. (2013) conceptualized a new document image descriptor by compressing and digitizing the descriptors. The method can be used with various public datasets to get better results.

Evans et al. (2015) conceptualized a method using event-related potentials (ERP) index for episodic and non-episodic tasks. They discussed critical reasons for positive results in one method compared to the other method. They conducted experiments to know the existence of the ERP for similar images.

Lin et al. [125] described a new method by integrating three different techniques for image retrieval. Statistical data from three color channels was extracted. The K-means algorithm and a classifier were used to create indexed data. They used natural image database for training and testing needs. Various comparisons were made in several ways, where good results were recorded in the process of recognition and retrieval.

In CBIR, a major issue to be handled is the existence of the noisy shapes and curves. Noise presents itself in the form of continuous distortions as well as the regional anatomy variations. It is well known that the persistent Betti numbers are very popular shape descriptors used to estimate the dissimilarities under continual distortions of the shape. In Patrizio and Claudia (2013), authors concentrated on the issue of noise which changes the structure of the objects under study. In this method, a general approach was proposed to transform the persistent Betti numbers into the steady descriptors even in the occurrence of structural variations. They tested their method on the Kimia-99 dataset to determine the efficiency of retrieval.

It is not possible to develop a processing system that can segregate different classes with properly identified boundaries in feature space. Hence, CBIR with traditional distance measure is not effective for image features of low level such as texture. For a classifier-based retrieval method, performance heavily relies on the performance of classifier. If categorization of query is correctly done, it exhibits high retrieval efficiency; whereas the total failure of retrieval system occurs if the misclassification is done. Hence, there is a large variation in the performance.

Sudipta et al. (2013) framed a novel method to CBIR to counter few drawbacks of traditional distance and classifier methods. In the first step, computation of the class label and the fuzzy membership was done. Subsequently, a combination of a simple and weighted distance metrics was adapted to develop a classifier-dependent retrieval framework. It is different from the conventional methods in terms of the search space considered. This method enables the process to minimize the search space in a different way to increase the speed of retrieval. However, the accuracy decreases gradually in this process. The effectiveness of the approach was estimated using three different texture datasets with different complexity levels, orientations and types of classes.

In commercial sectors, the trademarks are specific visual data with more ranking value, due to the quality perception and the amount of creativity involved. They are crucial prominent assets utilized as a business tool to depict the guarantee of the quality, innovation, and the standards maintained by the manufacturers. This necessitates the precaution for trademark protection by giving a solution to prevent infringement. The problem can be resolved by designing retrieval frameworks capable of identifying the visual trademark similarities.

Anuar et al. (2013) proposed an innovative framework for trademark retrieval by integrating global and local descriptors. Zernike moment's coefficients were treated as global descriptors while edge-gradient co-occurrence matrix was considered as local descriptor. The contour information plays a significant role in determining the visual similarity. Experiments conducted on MPEG-7 provided better retrieval results in terms of the standard precision.

In general, the memory retrieval involves restoration of cortical activity. Based on this philosophy, many studies on Functional Magnetic Resonance Imaging (fMRI) proved that there exists a relationship between brain regions during retrieval (Andrews-Hanna et al. (2014)). They used electroencephalogram to obtain retrieval success.

An algorithm to extract data in a more efficient manner was proposed in Thangaraj and Sujatha (2014). A set of plain keywords was given as the query input from user. Query was transformed into a query of semantics. The concepts from the domain ontologies and the third-party wordbook were used. The improved algorithm was adapted to extract the relevant information in retrieval and ranking. The experimental outcome proved that the improvement made to the algorithm helps to retrieve more relevant data from the web documents with good accuracy and efficiency.

2D cartoon has a significant place in so many fields. It needs efficient methods to avoid human labor. In Liang et al. (2013), a heterogeneous cartoon gesture recognition approach was proposed for different applications. In the first step, the diverse properties with varying dimensions were identified to indicate the cartoon and human-subject images as per their features. Subsequently, a framework was designed by combining shared structure learning (SSL) and graph-dependent learning to learn dependable classifiers on diverse characteristics. In a cross-feature manner, the framework was used to recognize and quantitatively evaluate the same similarity between the cartoon and human subject gestures. They conducted experiments extensively to highlight the effectiveness of the method and illustrated the application in a number of cartoon industry applications.

An increasing number of research works highlight human brain's involvement when the people recollect their past or mimic their future. The most advanced studies on heterogeneity inside a network increase the probability that these autobiographical systems include various processes backed by various anatomic subsystems.

Jessica et al. (2014) earlier proposed that biomedical subsystem adds to an autobiographical memory as well as future thinking by facilitating the people to retrieve the earlier information and compose the same. In contrast, dorsal subsystem was introduced back to social-reflective issues. The outcome of various experiments depict that default network is a heterogeneous brain system.

Control processes are important for expediting and suppressing the memory retrieval, but these systems are not well known. Motivated by a similar fMRI approach, Ketz et al. (2014) utilized a slightly changed Think/No-Think (TNT) pattern to inspect the neural impressions of purpose over suppressing and enhancing memory retrieval. Earlier methods showed memory improvement when the well-learned stimulus sets were reused in cued recall ("think or recall of analyzed set component") and deterioration when reused with cued suppression ("avert thinking of studied set component"). They applied division-based multivariate categorization of electroencephalography waves to actuate if respective target components are well suppressed or retrieved. The comparison results between controlled suppression and controlled retrieval showed highly prominent Theta oscillations in restrained retrieval. Beta oscillations captured in high levels of controlled suppression and control retrieval suggested for more usual control-relevant role.

To retrieve the architectural plans, Ahmed et al. (2014) devised a sketch-based and automatic analysis method. For testing a floor plan database, they proposed a SCatch method which was a sketch-based method. Then, a system was adapted to create a database. The second method explored semantic patterns drawn by an architect. They proposed new preprocessing techniques, e.g., the variation among thin, thick and medium lines as well as eviction of few components outside convex hull of external walls. The algorithm explored the semantic pattern drawn by the architect. It was established that the performance of the sketch recognition method was considerably high. The proposed technique can be used in practical applications also.

Mathieu et al. (2013) have conceptualized a novel exemplar-based technique for real-time motion recognition using Motion Capture (MoCap) information. They delineated streamed identifiable actions, emanating from an online MoCap engine, into a motion graph which is similar to an animated motion graph. In order to achieve better retrieval output on classification, the spatio-temporal metric was used in the analogy measurements. The proposed method had a positive edge of being linear and incremental, thereby executing the recognition process very fast.

Murala and Jonathan Wu (2014) devised an image retrieval method based on Local Mesh patterns for biomedical image retrieval. Probable relationships among nearby neighbors were decided by the number of nearby pixels, P. The proposed method was combined with Gabor transform to establish the efficiency of retrieval. A thorough analysis of the results from experiments proved the ability of proposed approach. It outperformed other techniques such as local binary pattern and local ternary patterns in terms of performance metrics.

To address the problems of region of interest (ROI) segmentation and image retrieval simultaneously, Ivan et al. (2014) introduced a probabilistic generative method. More specifically, the introduced method considered few attributes of the similarity among objects in different images. The accuracy in retrieved matches between any set of images was improved in the same way. The method also worked well simultaneously for segmentation of the regions of interest and objects of interest. By considering many objects of interest, this method was tested on three different databases: two of them were related to multi-object image retrieval, i.e. image segmentation and object detection, while the other one was related to multi-view image retrieval. These experiments highlighted the capacity of the method to consider the cases with more number of objects.

Concept of image retrieval was plenteously focused over the last two decades due to the ever increasing demands for the efficient use of multimedia data. A method called scale space representation and local key point descriptors was proposed by Park et al. (2014). After the introduction of Scale-Invariant Feature Transform, the idea of scale space realization gained wide acceptance as a powerful means to retrieve the images. Authors analyzed some properties of the scale space procedure and proposed an enhancement of the same approach which considerably increases the image matching veracity.

Content-based image retrieval systems usually blend relevance feedback method to improve retrieval results according to cognitive response from user. User provides feedback by identifying the related and unrelated images. Relevance decisions are usually considered to be category based. It is not suggested to put the user in the loop to decide the category of the image, even though the user is not familiar with the database. The best solution for this is to get the information in the form of relative similarity recognition. The interface which presents the retrieved images also plays a significant role (Han & McKenna 2014). Similarity-dependent 2D maps provide the context and enable more effective visual search. A collaborative browsing and retrieval depending upon similarity information derived from 2D image maps was highlighted. Online maximal margin learning was followed during the image resemblance measure in retrieval.

In recent times, content-based histology image retrieval exhibited high prospects in the field of image retrieval more specifically in hospital activities, learning, and biomedical research. Many studies proved that the combination of features has a significant role in increasing the discriminating capacity of visual properties related to semantically useful queries. It is particularly valuable in histology image analysis where inventive mechanisms are required to construe varying tissue configuration and architecture into histological methods.

Qianni et al. (2013) proposed an integrated method to combine diverse visual features for histology image retrieval. Based on the preferences of a group of query images, a multi-objective learning method to understand the optimal semantic matching was proposed in the work. The objective was to attain a better integrated model for a keyword related to many query images. The function was interpreted as an optimization issue, and an improved optimization methodology was used to handle critical contradictions.

Majid et al. (2013) made a study on image retrieval with machine learning and statistical resemblance matching methods. The objective of this survey was to use the power of texture and shape features of image in order to enhance the retrieving ability. Mechanism for segmenting the image into blocks of various dimensions, resulting in higher retrieval outcome, was proposed. In the presented method, the image was separated into blocks of different dimensions. The statistical parameters like standard deviation and energy were extracted. The extracted sub-features from each tile acts as a feature descriptor. In the next level, the shape feature was characterized by invariant edge moments. The method was tested on large-size image dataset to establish the better features such as high retrieval accuracy and low complexity.

Pan et al. (2014) introduced a modified scheme of Uncertain Location Graph for human brain's computed tomography image texture modeling. Upon modeling the texture image, ULG similarity extraction method was used and an efficient index pattern was applied to reduce the search time. A high accuracy and efficiency were noticed from the experiments.

Guo et al. (2010) proposed a less complex method named ordered-dither block truncation coding (ODBTC) to create a feature descriptor. A compression scheme was followed in the encoding stage. From ODBTC encrypted streams of image data, the features named Color Co-occurrence Feature (CCF) and Bit Pattern Features (BPF) were produced without any decoding stage. The improvement in experimental results when compared to BTC and other methods established that the method suits for compression as well as to create an efficient feature descriptor for image retrieval.

Xiaofan et al. (2015) proposed a supervised kernel hashing method to compress huge number of dimensional feature vectors into a very small number of binary digits while retaining the instructive signatures. Binary data was indexed into a hash table to facilitate retrieval from huge database. Specifically, supervised data was utilized to overpass the semantic difference in image features and symptomatic data. The method was validated by using thousands of images taken from microscopic tissues.

Georgios et al. (2014) devised one type of Relevance Feedback method for region-based retrieval. An iterative assessment of the user interested in real-time objects and exploitation of the same information to design a better retrieval system was followed in the work. Gaze features were derived to represent the prediction by the user and an efficient relevance feedback framework for retrieval system was designed. Further, the usage of suitable gaze tracker improves the cost of the entire mechanism and portability.

Chen et al. (2013) proposed a method to extract human attributes containing the cues of the facial images. Improvement in the retrieval performance is reported by extracting the code words related to the facial images. By taking the advantage of human attributes, they proposed orthogonal approaches called attribute enhanced sparse coding, attribute embedded inverted indexing to enhance online and offline facial image retrieval. They tested the potential of various attributes and critical elements needed in the face retrieval.

Yang et al. (2014) proposed a bag-of-object retrieval method to estimate the image relevance that was particularly effective for object queries. They created an object terminology possessing query-relevant objects by exploring the frequent object chunks from the resultant image collection. Upon creating a bag of objects for each image, the model was derived from a mechanism for a language modeling. They adapted a supervised framework to combine multiple ranking features and a two-step ranking method to outperform the other methods.

Daniela et al. (2013) proposed an Earth-observation (EO) image retrieval system using augmented metadata, semantic elucidation, and image content named EO retrieval. The EO retrieval produces an EO-data model by utilizing automatic feature extraction, treating the EO-created metadata, and delineating semantics, which was subsequently exploited to support complex queries. A catalog of TerraSAR-X E was used to illustrate the proposed method. The database contains 39 high-resolution scenes having nearly 50,000 patches, including the property descriptors, a hundred of metadata particulars for each scene, and about 330 semantic delineations.

Dubey et al. (2015) designed a descriptor using local diagonal extrema structures for computed tomography image retrieval. In this method, diagonal neighbors were considered to extract the spatial relationship among the pixels by taking the center pixel as reference. First-order derivatives were extracted based on the diagonal extrema scheme. The primary focus in the proposed work was to minimize the dimensionality problem most commonly faced in image retrieval. The method was tested on biomedical image databases such as NEMA-CT and Emphysema-CT to outline the efficiency of the system and a significant improvement was highlighted.

Verma and Raman (2016) proposed local tri-directional patterns for CBIR by using intensity of pixels taken from three directions in a neighborhood. Factor of magnitude was included to achieve improvement in the retrieval process.

Vipparthi et al. (2016) introduced a new approach to collect Maximum Edge and Position Patterns from magnitude directional edges. MEP and MEPP were calculated after extracting magnitude and position information from DE. The multi-resolution Gaussian filters were also used to increase the performance. Rao et al. (2013, 2014, 2016) proposed some image retrieval systems based on local patterns.

## 2.7  PROBLEM STATEMENT

A major step in CBIR systems is extraction of features. Among many features, texture provides more discriminating information within the image. A single pixel can't provide complete texture information with varying illumination, scale and rotation. However, correlation among neighboring pixels effectively represents texture information which is an important part of any retrieval system.

Local binary patterns explore the correlation among the pixels to extract the textural information. But, the size of feature vector increases with the increase of window size. Hence, the tradeoff between dimensionality of feature vector and computational accuracy greatly influences any practical system. The impact of extraction of significant information from the feature vector greatly reduces the computational complexity. Many researchers attempted to reduce the feature vector size. The challenge is to keep more amount of information while reducing the size of feature vector. This challenge has to be addressed from the standpoint of exploring the specificities of hidden information within the feature vector.

This book is aimed at addressing these challenges while maintaining the accuracy parameters like recall and precision in any content-based image retrieval system.

## 2.8  METHODOLOGY

Inherent texture information in an image can be effectively explored by considering correlation among a group of pixels rather than a single pixel. The neighborhood of pixels effectively depicts the structure of texture and thus is more useful in interpreting the content. Local binary patterns method was used to extract the texture from a selected kernel of varying sizes. It was meant for addressing the issues of rotational variance, scale, illumination. However, size of the feature vector becomes large when neighborhood size is increased. Based on the concept of local binary patterns, an attempt is made in the current work to increase accuracy of the retrieval system while maintaining less feature vector length.

Local extrema patterns are used to increase accuracy of retrieval system. Directional Local Extrema Patterns are explored to improve performance by adding color and Gabor features. Quantization is applied to minimize no. of pixels in a neighborhood under consideration. Color information is collected from two oppugnant planes to enhance efficiency of retrieval system. A mesh structure is formed as a variant of quantized extrema patterns. Five different databases with varying image content and dimensions are used to test the efficiency of the algorithms. Different types of distance measures such as Euclidean distance and D1 distance are used in the process. The MATLAB 7.0 is the platform used to evaluate the proposed algorithms and the system configuration is Intel core-i5, 2.3GHz. Evaluation is made in terms of precision, recall, average retrieval precision and average recall rate.

# 3 Improved Directional Local Extrema Patterns

## 3.1 INTRODUCTION

In this chapter, improvements made to Directional Local Extrema Patterns [DLEP] are presented. DLEP extract the spatial relationship among the pixels by considering the center pixel as a threshold. An attempt is made in the current work by adding color feature and Gabor feature to DLEP. The proposed methods are tested with Corel-1k database. Experiments showed considerable improved results in evaluation metrics.

## 3.2 LOCAL PATTERNS

### 3.2.1 LOCAL BINARY PATTERNS

Ojala et al. [76] proposed texture modeling scheme, in which the local image texture is described by gray-scale and rotational invariant local binary patterns (LBP) operator over a circular neighborhood as given in Figure 3.1. In LBP, global texture is described based on the discrete occurrence of local patterns extracted from a region of an image or entire image. Success of LBP is reported in some thrust zones of image processing. LBP is found in texture discrimination, texture segmentation and recognition of facial identity and expression.

Considering a pixel at the center in a 3×3 kernel, LBP are calculated by comparing intensity value with its surrounding elements as per Equations (3.1) and (3.2) given below

$$LBP_{S,T} = \sum_{a=1}^{S} 2^{(a-1)} \times g_1 \left( M\left(h_p\right) - M\left(h_c\right)\right) \tag{3.1}$$

$$g_1\left(b\right) = \begin{cases} 1 & b \geq 0 \\ 0 & else \end{cases} \tag{3.2}$$

where M(hc) represents intensity value of center pixel, M(hp) depicts intensity value of its corresponding neighbors, S denotes no. of surrounding picture elements and T shows radius of neighborhood.

Upon deriving pattern for every pixel $(y, z)$, entire image is depicted by constructing histogram using the equations specified below

$$Hist_{LBP}\left(m\right) = \sum_{y=1}^{L_1} \sum_{z=1}^{L_2} g_2\left(LBP\left(y,z\right),m\right); m \in \left[0,\left(2^S - 1\right)\right] \tag{3.3}$$

DOI: 10.1201/9781003123514-3

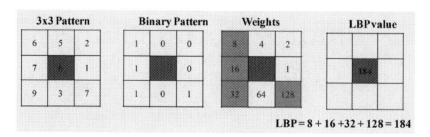

**FIGURE 3.1**  Illustration of local binary patterns.

$$g_2(c,d) = \begin{cases} 1 & c = d \\ 0 & else \end{cases} \tag{3.4}$$

where $L_1 \times L_2$ is the dimension of the image under consideration.

Figure 3.1 illustrates the LBP calculation for a given 3×3 pattern. The histograms which are built from the patterns provide information about presence of edges in image.

### 3.2.2  BLOCK-BASED LOCAL BINARY PATTERNS

Takala et al. (2005) introduced Block-Based Local Binary Patterns (BLK_LBP). This is computationally simple method which depends on sub-images to extract spatial features. This can be associated with histogram-based feature descriptors such as LBP. Initially, it partitions image into arbitrary-sized blocks. Upon dividing images, LBP are extracted for every block and histograms are built. The histograms of each sub-block are concatenated to create final form of vector for image retrieval.

### 3.2.3  CENTER-SYMMETRIC LOCAL BINARY PATTERNS

Heikkila et al. (2009) proposed a modified form of LBP named Center-Symmetric Local Binary Patterns (CS_LBP), wherein the center symmetric pixels are used for comparison purpose as per Equation (3.5) given below:

$$CS\_LBP_{S,T} = \sum_{a=1}^{S} 2^{(a-1)} \times g_1\left(M\left(h_p\right) - M\left(h_{p+(S/2)}\right)\right) \tag{3.5}$$

Similar to LBP, a histogram is built using the CS_LBP values for every pixel $(y,z)$.

### 3.2.4  LOCAL DIRECTIONAL PATTERN

Jabid et al. (2010) introduced Local Directional Patterns for human face recognition. Edge response of each pixel in multiple orientations is the primary concept of this method. Kirsch masks are applied to derive the spatial structure by exploring responses in eight orientations. Presence of edge or corner is indicated by high response values in the respective direction. Image is divided into regions to obtain data regarding spatial location. Followed by that, 56-bin histogram is built to depict the presence of edges.

## 3.3   DIRECTIONAL LOCAL EXTREMA PATTERNS

Based on the concept of LBP, Murala et al. (2012a) introduced DLEP for image retrieval. It extracts spatial composition of texture according to the local maxima or minima of center gray pixel. Rao et al. (2013, 2015) added certain features to DLEP to improve the performance.

Local maxima or minima in $0°$, $90°$, $45°$ and $135°$ are derived by calculating difference in center pixel intensity and of neighbors as specified here under.

$$M'(h_j) = M(h_c) - M(h_j); \quad j = 1, 2, \ldots, 8 \tag{3.6}$$

The local maxima or minima are calculated using Equation (3.7).

$$\hat{I}_\alpha(h_c) = g_3(I'(h_k), I'(h_{k+4})); \ k = (1 + \alpha/45) \forall \alpha = 0°, 45°, 90°, 135° \tag{3.7}$$

$$g_3(I'(h_k), I'(h_{k+4})) = \begin{cases} 1 & I'(h_k) \times I'(h_{k+4}) \geq 0 \\ 0 & else \end{cases} \tag{3.8}$$

DLEP is interpreted ($\alpha = 0°$, $45°$, $90°$, and $135°$) as follows:

$$DLEP(I(h_c))\big|_\alpha = \{\hat{I}_\alpha(h_c); \hat{I}_\alpha(h_1); \hat{I}_\alpha(h_2); \ldots \ldots \hat{I}_\alpha(h_8)\} \tag{3.9}$$

Procedure to compute local extrema in $45°$ is provided in Figure 3.2. If the intensity value of center pixel (yellow color) is either less than or more than both of its neighbors (green color) in that direction, a binary 1 is assigned, else 0 is assigned. Local extrema obtained for the image in Figure 3.2 becomes 011111100 and the same procedure can be followed to compute the extrema in $0°$, $90°$ and $135°$ directions. Table 3.1 gives the extrema patterns in the specified directions. Feature maps of directional local extrema are provided in Figure 3.3.

## 3.4   IMPROVED DIRECTIONAL LOCAL EXTREMA PATTERNS

Based on the concept of DLEP, we introduce two methods to improve the same: (i) combination of color and DLEP and (ii) combination of Gabor features and DLEP.

### 3.4.1   COMBINATION OF COLOR AND DLEP

In the proposed integrated approach, color features are collected using color moments and texture is explored by DLEP. Color moments show better discriminating ability

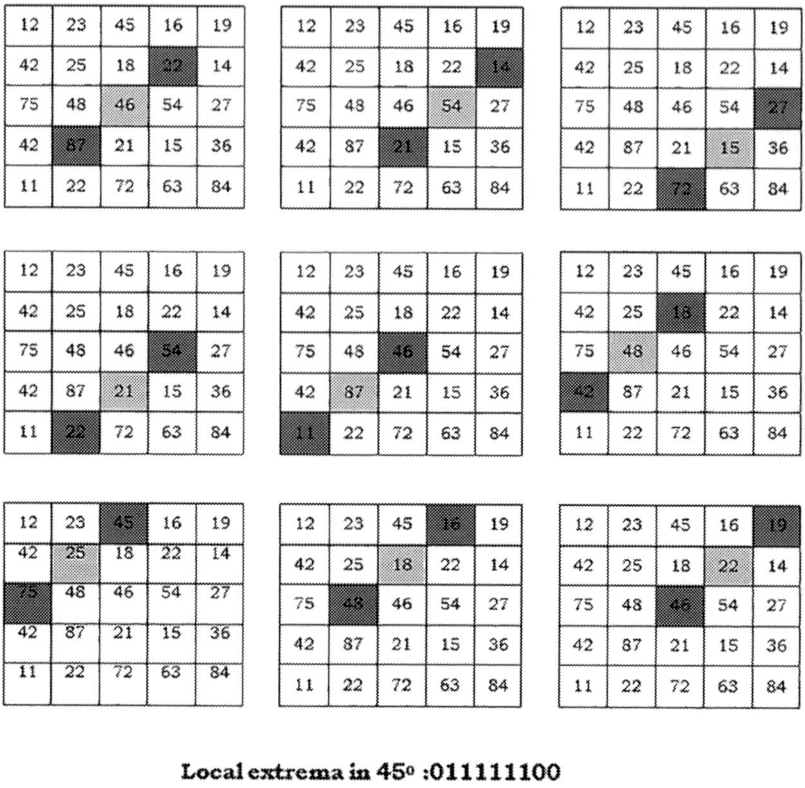

| 12 | 23 | 45 | 16 | 19 |
|----|----|----|----|----|
| 42 | 25 | 18 | 22 | 14 |
| 75 | 48 | 46 | 54 | 27 |
| 42 | 87 | 21 | 15 | 36 |
| 11 | 22 | 72 | 63 | 84 |

| 12 | 23 | 45 | 16 | 19 |
|----|----|----|----|----|
| 42 | 25 | 18 | 22 | 14 |
| 75 | 48 | 46 | 54 | 27 |
| 42 | 87 | 21 | 15 | 36 |
| 11 | 22 | 72 | 63 | 84 |

| 12 | 23 | 45 | 16 | 19 |
|----|----|----|----|----|
| 42 | 25 | 18 | 22 | 14 |
| 75 | 48 | 46 | 54 | 27 |
| 42 | 87 | 21 | 15 | 36 |
| 11 | 22 | 72 | 63 | 84 |

| 12 | 23 | 45 | 16 | 19 |
|----|----|----|----|----|
| 42 | 25 | 18 | 22 | 14 |
| 75 | 48 | 46 | 54 | 27 |
| 42 | 87 | 21 | 15 | 36 |
| 11 | 22 | 72 | 63 | 84 |

| 12 | 23 | 45 | 16 | 19 |
|----|----|----|----|----|
| 42 | 25 | 18 | 22 | 14 |
| 75 | 48 | 46 | 54 | 27 |
| 42 | 87 | 21 | 15 | 36 |
| 11 | 22 | 72 | 63 | 84 |

| 12 | 23 | 45 | 16 | 19 |
|----|----|----|----|----|
| 42 | 25 | 18 | 22 | 14 |
| 75 | 48 | 46 | 54 | 27 |
| 42 | 87 | 21 | 15 | 36 |
| 11 | 22 | 72 | 63 | 84 |

| 12 | 23 | 45 | 16 | 19 |
|----|----|----|----|----|
| 42 | 25 | 18 | 22 | 14 |
| 75 | 48 | 46 | 54 | 27 |
| 42 | 87 | 21 | 15 | 36 |
| 11 | 22 | 72 | 63 | 84 |

| 12 | 23 | 45 | 16 | 19 |
|----|----|----|----|----|
| 42 | 25 | 18 | 22 | 14 |
| 75 | 48 | 46 | 54 | 27 |
| 42 | 87 | 21 | 15 | 36 |
| 11 | 22 | 72 | 63 | 84 |

| 12 | 23 | 45 | 16 | 19 |
|----|----|----|----|----|
| 42 | 25 | 18 | 22 | 14 |
| 75 | 48 | 46 | 54 | 27 |
| 42 | 87 | 21 | 15 | 36 |
| 11 | 22 | 72 | 63 | 84 |

**Local extrema in 45° :011111100**

FIGURE 3.2    Illustration of local extrema calculation.

TABLE 3.1
## Calculation of Directional Local Extrema

| Angle | Position | | | | | | | | | |
|-------|---|---|---|---|---|---|---|---|---|------|
|       | 0 | 1 | 2 | 3 | 4 | 5 | 6 | 7 | 8 | DLEP |
| 0°    | 0 | 0 | 0 | 0 | 1 | 1 | 0 | 1 | 0 | 467 |
| 45°   | 0 | 1 | 1 | 1 | 1 | 1 | 1 | 0 | 0 | 252 |
| 90°   | 1 | 1 | 0 | 0 | 1 | 1 | 1 | 1 | 0 | 498 |
| 135°  | 1 | 1 | 0 | 0 | 0 | 1 | 1 | 1 | 0 | 511 |

in identifying similar color images. In current research work, the mean, standard deviation and skewness in Rao and Venkata Rao (2014a) are considered to explore color feature for a RGB image.

Corel-1k database is used to estimate the performance. Experiments show that improved framework achieves better values in precision and recall when compared to the existing DLEP. The algorithm is provided below and the schematic is provided in Figure 3.4.

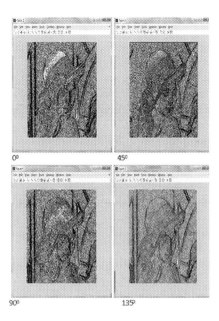

**FIGURE 3.3** Directional feature maps in 0°, 45°, 90°, 135°.

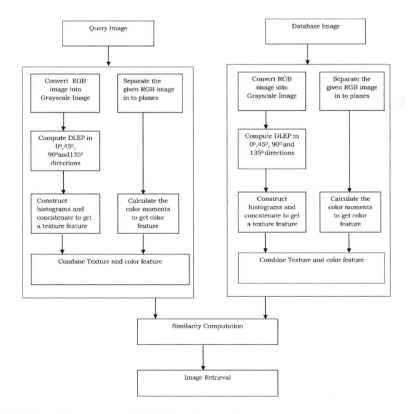

**FIGURE 3.4** Framework of CDLEP-based retrieval system.

**ALGORITHM**

1. Convert RGB image into gray-scale image.
2. Derive local maxima or minima in 0°, 45°, 90° and 135° directions.
3. Compose patterns in four directions as specified in step 2.
4. Build histogram of extrema patterns and concatenate all four to obtain texture feature vector.
5. Calculate color moments in R, G, B planes to create feature color feature descriptor.
6. Integrate the color and texture features to create CDLEP, i.e. Color Directional Local Extrema Pattern.
7. Compare query and database images.
8. Perform retrieval based on similarity level.

*Proposed Framework*

Each image is treated as a query during the experimentation. Subsequently, $d_1$ distance measure mentioned in Equation 1.3 is applied to estimate the similarity value for the image.

*Experimental Results and Discussions*

To test performance of proposed approach, Corel-1k database is considered. Evaluation is done in terms of precision and recall. Precision and recall values are provided in Tables 3.2 and 3.3, respectively. Figures 3.5 and 3.6 provide the comparison of proposed approach with the DLEP method. The retrieval results are shown in Figure 3.7 with a query at the top left.

---

**TABLE 3.2**
**Retrieval results of CDLEP and DLEP in Terms of Precision on Corel-1k**

| Category | DLEP | CDLEP |
|---|---|---|
| Africans | 69.3 | 71.5 |
| Beach | 60.5 | 63.5 |
| Building | 72.0 | 72.5 |
| Buses | 97.9 | 90 |
| Dinosaur | 98.5 | 99.2 |
| Elephant | 55.9 | 54.2 |
| Flower | 91.9 | 92.6 |
| Horse | 76.9 | 79.2 |
| Mountain | 42.7 | 45.8 |
| Food | 82.0 | 83.6 |
| **Average Precision (%)** | **74.8** | **75.2** |

---

**TABLE 3.3**

**Retrieval Results of CDLEP (DLEP+Color) and DLEP in Terms of Recall on Corel-1k.**

| Category | DLEP | CDLEP |
|----------|------|-------|
| Africans | 39.7 | 41 |
| Beach | 37.3 | 39 |
| Building | 34.9 | 41 |
| Buses | 74.1 | 67 |
| Dinosaur | 88.0 | 85 |
| Elephant | 29.0 | 34 |
| Flower | 70.8 | 83 |
| Horse | 41.7 | 42 |
| Mountain | 29.0 | 30 |
| Food | 47.0 | 43 |
| **Average Recall (%)** | **49.16** | **50.5** |

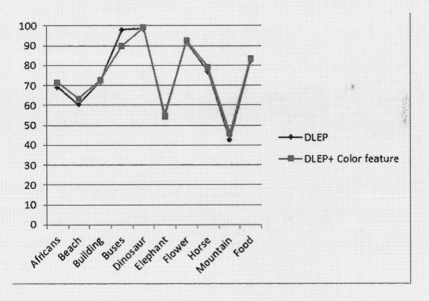

**FIGURE 3.5**   Comparison of precision values for DLEP and CDLEP.

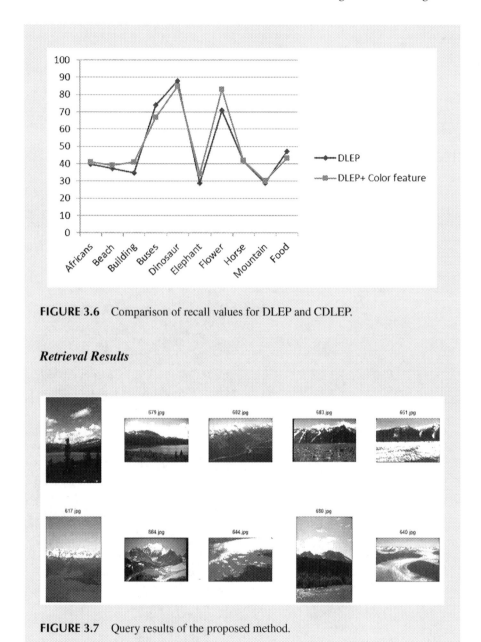

**FIGURE 3.6**   Comparison of recall values for DLEP and CDLEP.

*Retrieval Results*

**FIGURE 3.7**   Query results of the proposed method.

### 3.4.2   COMBINATION OF DLEP AND GABOR FEATURES

In the proposed integrated approach, Gabor features are added to DLEP features. The method is tested on standard database Corel-1k. Results obtained in the improved model and DLEP are shown in Tables 3.4 and 3.5. The algorithm and diagram for the

**TABLE 3.4**
**Retrieval Results of GDLEP, CDLEP and DLEP in Terms of Precision on Corel-1k**

| Category | Precision (%) DLEP | CDLEP | GDLEP |
|---|---|---|---|
| Africans | 69.3 | 71.5 | 73.5 |
| Beach | 60.5 | 63.5 | 62.5 |
| Building | 72.0 | 72.5 | 74.9 |
| Buses | 97.9 | 90 | 90 |
| Dinosaur | 98.5 | 99.2 | 98.7 |
| Elephant | 55.9 | 54.2 | 57.5 |
| Flower | 91.9 | 92.6 | 92 |
| Horse | 76.9 | 79.2 | 79.4 |
| Mountain | 42.7 | 45.8 | 47.8 |
| Food | 82.0 | 83.6 | 83 |
| Average Precision (%) | 74.8 | 75.2 | 75.9 |

**TABLE 3.5**
**Retrieval Results of GDLEP, CDLEP and DLEP in Terms of Recall**

| Category | DLEP | DLEP + Color Feature | DLEP + Gabor Feature |
|---|---|---|---|
| Africans | 39.7 | 41 | 40.5 |
| Beach | 37.3 | 39 | 38.4 |
| Building | 34.9 | 41 | 36.5 |
| Buses | 74.1 | 67 | 67 |
| Dinosaur | 88 | 85 | 84.6 |
| Elephant | 29 | 34 | 33.6 |
| Flower | 70.8 | 83 | 82.5 |
| Horse | 41.7 | 42 | 42 |
| Mountain | 29 | 30 | 30.5 |
| Food | 47 | 43 | 44.6 |
| Average Recall (%) | 49.1 | 50.5 | 50.0 |

GDLEP (Figure 3.8) are provided below. Experimental results show that the proposed approach achieves increased precision and recall in contrast to DLEP as shown in Figures 3.9 and 3.10. The query results are provided in Figure 3.9.

From the above data, the following observations are made.

1. Average precision of CDLEP (75.2%) is slightly more when compared to the existing DLEP (74.8%).
2. Average recall of CDLEP (50.5%) is slightly more when compared to DLEP (49.1%).
3. Average precision of GDLEP (75.9%) is slightly more when compared to the existing DLEP (74.8%).
4. Average recall of GDLEP (50.0%) is slightly more when compared to DLEP (49.1%)

**ALGORITHM**

1. Convert RGB image into gray-scale image.
2. Compute Gabor response for given image in three scales and eight directions.
3. Calculate local maxima or minima in 0°, 45°, 90° and 135°.
4. Compose extrema patterns in four directions.
5. Construct histograms local extrema patterns and concatenate.
6. Integrate two features in Step 2 and 5 to form feature descriptor.

*Framework of GDLEP*

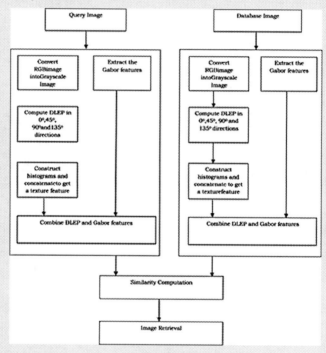

**FIGURE 3.8**    Framework of proposed GDLEP-based retrieval system

*Experimental Results and Discussions*

Corel-1k is employed to test the effectiveness of this technique. Results are provided in Tables 3.4 and 3.5. Comparison is given in Figures 3.9 and 3.10. Figure 3.11 depicts the retrieval results.

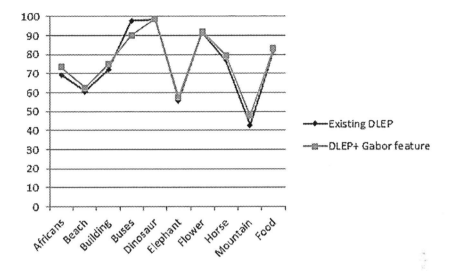

**FIGURE 3.9**    Comparison of precision values for DLEP and GDLEP.

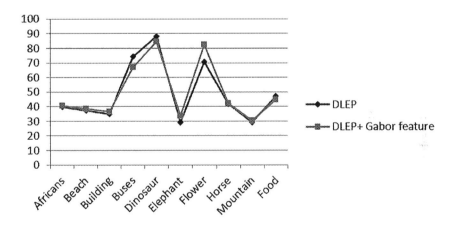

**FIGURE 3.10**    Comparison of recall values for DLEP and GDLEP.

**FIGURE 3.11**    Query results of the proposed approach.

## 3.5   CONCLUSION

In this chapter, improved models of DLEP are discussed. In CDLEP method, the color feature is extracted using color moments and combined with texture feature in the form of DLEP. Effectiveness of the proposed approach is tested on standard database, i.e. Corel-1k. The proposed method exhibits slight improvement in the recall values and the precision values.

In the second method, the Gabor features are added to DLEP to get improvement in the retrieval performance. The proposed GDLEP shows improvement in precision and recall values when compared to DLEP. However, the improvement is observed only in precision values when compared to CDLEP.

A novel feature descriptor named Local Quantized extrema patterns is presented in Chapter 4. It is aimed at achieving the better precision and recall values than DLEP and other methods.

### SOLVED PROBLEMS

1. Calculate LBP for the image of 3 × 3 size.

| 123 | 56 | 90 |
|-----|----|----|
| 234 | 80 | 66 |
| 39  | 78 | 62 |

**Solution:**

| 123 | 56 | 90 |
|-----|----|----|
| 234 | 80 | 66 |
| 39  | 78 | 62 |

$\Rightarrow$

| 1 | 0 | 1 |
|---|---|---|
| 1 |   | 0 |
| 0 | 0 | 0 |

$= 1\ 0\ 1\ 0\ 0\ 0\ 0\ 1$

$= 1 \times 2^7 + 0 \times 2^6 + 1 \times 2^5 + 0 \times 2^4 + 0 \times 2^3 + 0 \times 2^2$
$\quad + 0 \times 2^1 + 1 \times 2^0$

$= 127 + 0 + 32 + 0 + 0 + 0 + 0 + 1$

LBP Code = 160

Here, the threshold value is 80 (center pixel) $(g_c)$

IF $g_p \geq g_c = 1$
$g_p < g_c = 0$

where $g_p$ is neighbor pixel.

2. For a given 5×5 size, calculate LBP for C (2,3)

| 0 | 0 | 1 | 1 | 1 |
|---|---|---|---|---|
| 0 | 0 | 1 | 1 | 1 |  C (2,3)
| 0 | 2 | 2 | 2 | 2 |
| 2 | 2 | 3 | 3 | 3 |
| 2 | 2 | 3 | 3 | 3 |

**Solution:**

| | 0 | 1 | 2 | 3 | 4 |
|---|---|---|---|---|---|
| 0 | 0 | 0 | 1 | 1 | 1 |
| 1 | 0 | 0 | 1 | 1 | 1 |
| 2 | 0 | 2 | 2 | 2 | 2 |
| 3 | 2 | 2 | 3 | 3 | 3 |
| 4 | 2 | 2 | 3 | 3 | 3 |

→ Y

↓ X

C (2,3) $3 \times 3$ pixel

| | | |
|---|---|---|
| 1 | 1 | 1 |
| 2 | 2 | 2 |
| 3 | 3 | 3 |

| | | |
|---|---|---|
| 0 | 0 | 0 |
| 1 | | 1 |
| 1 | 1 | 1 |

Here center pixel $g_c$ is 2.

$$= 0\,0\,0\,1\,1\,1\,1\,1$$
$$= 0 \times 2^7 + 0 \times 2^6 + 0 \times 2^5 + 1 \times 2^4 + 1 \times 2^3 + 1 \times 2^2 + 1 \times 2^1 + 1 \times 2^0$$
$$= 0 + 0 + 0 + 16 + 8 + 4 + 2 + 1$$

LBP Code = 31

3. For the image of 3×3 size, find out whether it is having uniform LBP patterns or not.

| | | |
|---|---|---|
| 0 | 7 | 4 |
| 5 | 5 | 1 |
| 3 | 6 | 1 |

**Solution:**

From the given 3×3 image size, center pixel is $g_c = 5$.

| | | |
|---|---|---|
| 0 | 7 | 4 |
| 5 | 5 | 1 |
| 3 | 6 | 1 |

⇉

| | | |
|---|---|---|
| 0 | 1 | 0 |
| 1 | | 0 |
| 0 | 1 | 0 |

= 0 1 0 0 0 1 0 1

From the obtained binary pattern

0 1 0 0 0 1 0 1

1 2 3 4 5

There are total five transitions [i.e. 0→1, 1→0, 0→1, 1→0, 0→1]. Therefore, given 3×3 image size is having non-uniform LBP pattern.

4. For the image of 5×5 size, find LBP with gray-scale variance at a center of (3,3).

| 1 | 1 | 2 | 2 | 2 |
|---|---|---|---|---|
| 1 | 1 | 2 | 2 | 2 |
| 1 | 3 | 3 | 3 | 3 |
| 4 | 4 | 4 | 5 | 5 |
| 4 | 4 | 5 | 5 | 5 |

**Solution:**

LBP with gray-scale variance can be found

$$LBP_{(P,R)} = \sum_{P=0}^{P-1} S(g_p - g_c) 2^P$$

Where $S(x) = \begin{cases} 1, & x \geq 0 \\ 0, & x < 0 \end{cases}$

$g_p$ neigbor pixel, $g_c$ center pixel

|   | 0 | 1 | 2 | 3 | 4 |
|---|---|---|---|---|---|
| 0 | 1 | 1 | 2 | 2 | 2 |
| 1 | 1 | 1 | 2 | 2 | 2 |
| 2 | 1 | 3 | 3 | 3 | 3 |
| 3 | 4 | 4 | 4 | 5 | 5 |
| 4 | 4 | 4 | 5 | 5 | 5 |

| 3 | 3 | 3 |
|---|---|---|
| 4 | 5 | 5 |
| 5 | 5 | 5 |

$\Rightarrow$

| 0 | 0 | 0 |
|---|---|---|
| 0 |   | 1 |
| 1 | 1 | 1 |

$$LBP_{(P,R)} = \sum_{P=0}^{P-1} S(g_p - g_c) 2^P = LBP_{(P,R)} = \sum_{P=0}^{7} S(g_p - g_c) 2^P$$

$= 0\,(3{-}5)\,2^0 + 0\,(\,3{-}5)\,2^1 + 0\,(3{-}5)\,2^2 + 1\,(5{-}5)\,2^3 + 1\,(5{-}5)\,2^4 + 1\,(5{-}5)\,2^5$
$+ 1\,(5{-}5)\,2^6 + 0\,(4{-}5)\,2^7$

$= 0 + 0 + 0 + 0 + 0 + 0 + 0 + 0$

LBP (P,R) = 0

5. Calculate LBP and histogram for the given 5 × 5 kernel.

| 230 | 229 | 232 | 234 | 235 |
|-----|-----|-----|-----|-----|
| 237 | 236 | 236 | 234 | 233 |
| 255 | 255 | 255 | 251 | 230 |
| 99 | 90 | 67 | 37 | 94 |
| 222 | 152 | 255 | 129 | 129 |

| 0 | 0 | 0 | 0 | 0 | 0 | 0 |
|---|---|---|---|---|---|---|
| 0 | 230 | 229 | 232 | 234 | 235 | 0 |
| 0 | 237 | 236 | 236 | 234 | 233 | 0 |
| 0 | 255 | 255 | 255 | 251 | 230 | 0 |
| 0 | 99 | 90 | 67 | 37 | 94 | 0 |
| 0 | 222 | 152 | 255 | 129 | 129 | 0 |
| 0 | 0 | 0 | 0 | 0 | 0 | 0 |

**Solution:**

LBP Calculation for the 1st Row Elements

1.

| 0 | 0 | 0 |
|---|-----|-----|
| 0 | 230 | 229 |
| 0 | 237 | 236 |

| 0 | 0 | 0 |
|---|---|---|
| 0 | | 0 |
| 0 | 1 | 1 |

$\rightarrow$ 0000 1100
$= 0 \times 2^7 + 0 \times 2^6 + 0 \times 2^5 + 0 \times 2^4 + 1 \times 2^3 + 1 \times 2^2$
$+ 0 \times 2^1 + 0 \times 2^0$
$= 0+0+0+0+8+4+0+0 = 12$
LBP = 12

2.

| 0 | 0 | 0 |
|-----|-----|-----|
| 230 | 229 | 232 |
| 237 | 236 | 236 |

| 0 | 0 | 0 |
|---|---|---|
| 1 | | 1 |
| 1 | 1 | 1 |

$\rightarrow$ 0001 1111
$= 0+0+0+2^4+2^3+2^2+2^1+2^0 = 31$
LBP = 31

3.

| 0 | 0 | 0 |
|-----|-----|-----|
| 229 | 232 | 234 |
| 236 | 236 | 234 |

| 0 | 0 | 0 |
|---|---|---|
| 0 | | 1 |
| 1 | 1 | 1 |

$\rightarrow$ 0001 1110
$= 0+0+0+2^4+2^3+2^2+2^1+0 = 30$
LBP = 30

4.

| 0 | 0 | 0 |
|-----|-----|-----|
| 232 | 234 | 235 |
| 236 | 234 | 233 |

| 0 | 0 | 0 |
|---|---|---|
| 0 | | 1 |
| 1 | 1 | 0 |

$\rightarrow$ 0001 0110
$= 0+0+0+2^4+0+2^2+2^1+0 = 22$
LBP = 22

5.

| 0 | 0 | 0 |
|-----|-----|---|
| 234 | 235 | 0 |
| 234 | 233 | 0 |

| 0 | 0 | 0 |
|---|---|---|
| 0 | | 0 |
| 0 | 0 | 0 |

$\rightarrow$ 0000 0000
$= 0+0+0+0+0+0+0+0 = 0$
LBP=0

LBP Calculation for the 2nd Row Elements

6.

| 0 | 230 | 229 |
|---|-----|-----|
| 0 | 237 | 236 |
| 0 | 255 | 255 |

| 0 | 0 | 0 |
|---|---|---|
| 0 | | 0 |
| 1 | 1 | 1 |

$\rightarrow$ 0000 1110
$= 0+0+0+0+1 \times 2^3 + 1 \times 2^2$
$+1 \times 2^1 + 0 = 14$
LBP = 14

7.

| 230 | 229 | 232 |
|-----|-----|-----|
| 237 | 236 | 236 |
| 255 | 255 | 255 |

| 0 | 0 | 0 |
|---|---|---|
| 1 | | 1 |
| 1 | 1 | 1 |

$\rightarrow$ 0001 1111
$= 0+0+0+2^4+2^3+2^2+2^1+2^0 = 31$
LBP = 31

8.

| 229 | 232 | 234 |
|-----|-----|-----|
| 236 | 236 | 234 |
| 255 | 255 | 251 |

| 0 | 0 | 0 |
|---|---|---|
| 1 | | 0 |
| 1 | 1 | 1 |

$\rightarrow$ 0000 1111
$= 0+0+0+0+2^3+2^2+2^1+2^0 = 16$
LBP = 16

9.

| 232 | 234 | 235 |
|-----|-----|-----|
| 236 | 234 | 233 |
| 255 | 251 | 230 |

| 0 | 1 | 1 |
|---|---|---|
| 1 | | 0 |
| 1 | 1 | 0 |

$\rightarrow$ 0110 0111
$= 0+2^6+2^5+0+0+2^2+2^1+2^0 = 103$
LBP = 103

10.

| 234 | 235 | 0 |
|-----|-----|---|
| 234 | 233 | 0 |
| 251 | 230 | 0 |

| 1 | 1 | 0 |
|---|---|---|
| 1 |   | 0 |
| 1 | 0 | 0 |

→  1100 0011
$= 2^7+2^6+0+0+0+2^1+2^0 = 175$
LBP = 175

11.

| 0 | 237 | 236 |
|---|-----|-----|
| 0 | 255 | 255 |
| 0 | 99  | 90  |

| 0 | 0 | 0 |
|---|---|---|
| 0 |   | 1 |
| 0 | 0 | 0 |

→  0001 0000
$= 0+0+0+2^4+0+0+0+0 = 16$
LBP = 16

12.

| 237 | 236 | 236 |
|-----|-----|-----|
| 255 | 255 | 255 |
| 99  | 90  | 67  |

| 0 | 0 | 0 |
|---|---|---|
| 1 |   | 1 |
| 0 | 0 | 0 |

→  0001 0001
$= 0+0+0+2^4+0+0+0+2^0 = 17$
LBP = 17

13.

| 236 | 236 | 234 |
|-----|-----|-----|
| 255 | 255 | 251 |
| 90  | 67  | 37  |

| 0 | 0 | 0 |
|---|---|---|
| 1 |   | 0 |
| 0 | 0 | 0 |

→  0000 0001
$= 0+0+0+0+0+0+0+2^0 = 1$
LBP = 1

14.

| 236 | 344 | 233 |
|-----|-----|-----|
| 255 | 251 | 230 |
| 90  | 67  | 37  |

| 0 | 0 | 0 |
|---|---|---|
| 1 |   | 0 |
| 0 | 0 | 0 |

→  0000 0001
$= 0+0+0+0+0+0+0+2^0 = 1$
LBP = 1

15.

| 344 | 233 | 0 |
|-----|-----|---|
| 251 | 230 | 0 |
| 37  | 94  | 0 |

| 1 | 1 | 0 |
|---|---|---|
| 1 |   | 0 |
| 0 | 0 | 0 |

→  1100 0001
$= 2^7+2^6+0+0+0+0+0+2^0 = 193$
LBP = 193

16.

| 0 | 255 | 255 |
|---|-----|-----|
| 0 | 99  | 90  |
| 0 | 222 | 152 |

| 0 | 1 | 1 |
|---|---|---|
| 0 |   | 0 |
| 0 | 1 | 1 |

→  0110 1100
$= 0+2^6+2^5+0+2^3+2^2+0+0 = 108$
LBP = 108

17.

| 255 | 255 | 255 |
|-----|-----|-----|
| 99  | 90  | 67  |
| 222 | 152 | 255 |

| 1 | 1 | 1 |
|---|---|---|
| 1 |   | 0 |
| 1 | 1 | 1 |

→  1110 1111
$= 2^7+2^6+2^5+0+2^3+2^2+2^1+2^0 = 239$
LBP = 239

18.

| 255 | 255 | 251 |
|-----|-----|-----|
| 90  | 67  | 37  |
| 152 | 255 | 129 |

| 1 | 1 | 1 |
|---|---|---|
| 1 |   | 0 |
| 1 | 1 | 1 |

→  1110 1111
$= 2^7+2^6+2^5+0+2^3+2^2+2^1+2^0 = 239$
LBP = 239

19.

| 255 | 251 | 230 |
|-----|-----|-----|
| 67 | 37 | 94 |
| 255 | 129 | 129 |

| 1 | 1 | 1 |
|---|---|---|
| 1 | | 1 |
| 1 | 1 | 1 |

$\rightarrow$ 1111 1111
$=2^7+2^6+2^5+2^4+2^3+2^2+2^1+2^0 =$
255
LBP= 255

20.

| 251 | 230 | 0 |
|-----|-----|---|
| 37 | 94 | 0 |
| 129 | 129 | 0 |

| 1 | 1 | 0 |
|---|---|---|
| 0 | | 0 |
| 1 | 1 | 0 |

$\rightarrow$ 1100 0110
$= 2^7+2^6+0+0+0+2^2+2^1+0 =$
198
LBP = 198

## LBP Calculation for the 5th Row Elements

21.

| 0 | 99 | 90 |
|---|----|----|
| 0 | 222 | 152 |
| 0 | 0 | 0 |

| 0 | 0 | 0 |
|---|---|---|
| 0 | | 0 |
| 0 | 0 | 0 |

$\rightarrow$ 0000 0000
$= 0+0+0+0+0+0+0+0 =0$
LBP = 0

22.

| 99 | 90 | 67 |
|----|----|----|
| 222 | 152 | 255 |
| 0 | 0 | 0 |

| 0 | 0 | 0 |
|---|---|---|
| 1 | | 1 |
| 0 | 0 | 0 |

$\rightarrow$ 0001 0001
$= 0+0+0+2^4+0+0+0+2^0 = 17$
LBP = 17

23.

| 90 | 67 | 37 |
|----|----|----|
| 152 | 255 | 129 |
| 0 | 0 | 0 |

| 0 | 0 | 0 |
|---|---|---|
| 0 | | 0 |
| 0 | 0 | 0 |

$\rightarrow$ 0000 0000
$= 0+0+0+0+0+0+0+0 = 0$
LBP = 0

24.

| 67 | 37 | 94 |
|----|----|----|
| 255 | 129 | 129 |
| 0 | 0 | 0 |

| 0 | 0 | 0 |
|---|---|---|
| 1 | | 1 |
| 0 | 0 | 0 |

$\rightarrow$ 0001 0001
$= 0+0+0+2^4+0+0+0+2^0 = 17$
LBP = 17

25.

| 37 | 94 | 0 |
|----|----|---|
| 129 | 129 | 0 |
| 0 | 0 | 0 |

| 0 | 0 | 0 |
|---|---|---|
| 1 | | 0 |
| 0 | 0 | 0 |

$\rightarrow$ 0000 0001
$= 0+0+0+0+0+0+0+2^0 = 1$
LBP=1

The 5 × 5 matrix after calculating LBPs is given by

| 12 | 62 | 60 | 22 | 0 |
|----|----|----|-----|-----|
| 12 | 62 | 23 | 103 | 195 |
| 16 | 239 | 239 | 255 | 198 |
| 0 | 17 | 0 | 17 | 1 |

| 0 -> 3 | 60 -> 1 |
|--------|---------|
| 1 -> 3 | 62 - >2 |
| 12 -> 2 | 103->1 |
| 16 -> 1 | 108 -> 1 |
| 17 -> 3 | 193 -> 1 |

| 22 -> 1 | 195 -> 1 |
|---------|----------|
| 23 -> 1 | 198 -> 1 |
|         | 239 ->2  |
|         | 255 ->1  |

## HISTOGRAM

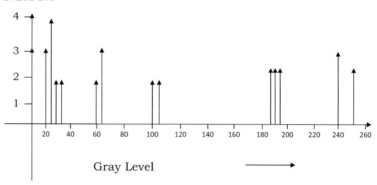

Gray Level

## EXERCISES

1.  Calculate LBP for the image of 3 × 3 size.

| 82 | 41 | 27 |
|----|----|----|
| 12 | 60 | 34 |
| 93 | 55 | 28 |

2.  For the image of 5 × 5 size, calculate LBP for C (3,3)

| 11 | 2  | 48 | 65 | 15 |
|----|----|----|----|----|
| 23 | 50 | 31 | 15 | 13 |
| 8  | 28 | 62 | 95 | 12 |
| 21 | 71 | 13 | 42 | 87 |
| 14 | 52 | 38 | 6  | 78 |

C (3,3)

3. For the image of 7 × 7 size, calculate LBP for C (5,6)

| 21 | 44 | 33 | 13 | 36 | 77 | 25 |
|----|----|----|----|----|----|----|
| 61 | 99 | 28 | 75 | 45 | 89 | 33 |
| 31 | 78 | 56 | 83 | 29 | 51 | 26 |
| 54 | 69 | 18 | 22 | 90 | 13 | 45 |
| 63 | 17 | 80 | 99 | 47 | 66 | 92 |
| 36 | 81 | 76 | 10 | 52 | 22 | 15 |
| 12 | 37 | 21 | 19 | 64 | 27 | 63 |

4. For the image of 3 × 3 size, find out whether it is having uniform LBP patterns or not

| 32 | 9  | 42 |
|----|----|----|
| 8  | 15 | 11 |
| 23 | 14 | 21 |

5. For the image of 5 × 5 size, find LBP with gray-scale variance at a center of (2,3)

| 21 | 14 | 2  | 58 | 2  |
|----|----|----|----|----|
| 18 | 22 | 6  | 24 | 2  |
| 82 | 37 | 76 | 31 | 43 |
| 41 | 94 | 24 | 56 | 45 |
| 53 | 26 | 75 | 7  | 90 |

# 4 Local Quantized Extrema Patterns

## 4.1 INTRODUCTION

This chapter provides the design of a texture descriptor, local quantized extrema patterns (LQEP), which adapts the concept of LQP (Hussain and Triggs 2012) and local directional extrema patterns (DLEP) (Murala et al. 2012a). The local binary pattern (LBP) operator extracts the spatial arrangement of texture by computing the gray-level differences. DLEP method considers all neighbors during the process of feature vector generation. Proposed LQEP approach extracts textural structure as per the variation between referenced picture element and the neighbors in four directions. Performance of proposed LQEP is evaluated with standard databases such as Corel-1k, Corel-5k and MIT VisTex. It is observed that proposed approach exhibits considerable improvement in precision and recall values as compared to recent techniques.

### 4.1.1 LOCAL QUANTIZED PATTERNS

Hussain and Triggs (2012) conceptualized local quantized patterns (LQP) for visual perception. LQP extract directional geometric properties in horizontal (H), vertical (V), diagonal (D) and anti-diagonal (A) strips of pixels; combinations of these such as horizontal-vertical (HV), diagonal-anti-diagonal (DA) and horizontal vertical-diagonal-anti diagonal (HVDA); and conventional disk-shaped and circular zones. Figure 4.1 exemplifies possible orientations of geometrically quantized structures for LQP operator. A detailed analysis of LQP is available in Hussain and Triggs (2012).

## 4.2 LOCAL QUANTIZED EXTREMA PATTERNS

Directional Local Extrema Patterns (Murala et al. 2012b) and LQP-based systems (Hussain and Triggs 2012) motivated us to introduce local quantized extrema patterns (LQEP). Proposed method in L. K. Rao et al. (2015) integrates local quantized patterns and directional local extrema patterns. Initially, possible structures are extracted from given image data. Subsequently, extrema values are collected from directional geometric structures. For a given 7×7 pattern, computation of LQEP operator is illustrated in Figure 4.2. For better understanding, 7×7 pattern in Figure 4.2 is indexed with pixel positions instead of intensity values. In current work, $HVDA_7$ geometric structure is considered to extract feature. Procedure to extract LQEP operator is given as follows.

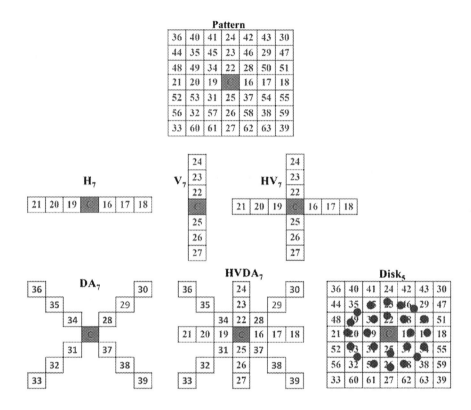

**FIGURE 4.1**    Possible geometric quantized structures.

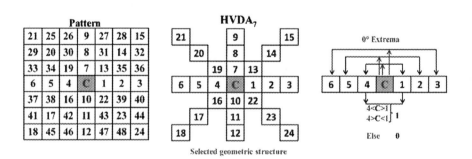

**FIGURE 4.2**    Illustration of LQEP for a 7×7 pattern.

For a specified pixel (P) at center of image $M$, $HVDA_7$ quantized geometric structure is extracted as shown in Figure 4.2. Then, maxima or minima (extrema) in four
directions, specifically in $0°$, $45°$, $90°$ and $135°$, are derived as follows:

$$DLE(M(x_P))\big|_{0°}$$
$$= \left\{ b_n(M(x_1), M(x_4), M(x_P)); b_n(M(x_2), M(x_5), M(x_P)); b_n(M(x_3), M(x_6), M(x_P)) \right\}$$

$$(4.1)$$

$$DLE(M(x_P))\big|_{45°}$$
$$= \{b_n(M(x_{13}),M(x_{16}),M(x_P));b_n(M(x_{14}),M(x_{17}),M(x_P));b_n((x_{15}),M(x_{18}),M(x_P))\}$$

$$(4.2)$$

$$DLE(M(x_P))\big|_{90°}$$
$$= \{b_n(M(x_7),M(x_{10}),M(x_P));b_n(M(x_8),M(x_{11}),M(x_P));b_n(M(x_9),M(x_{12}),M(x_P))\}$$

$$(4.3)$$

$$DLE(M(x_P))\big|_{135°}$$
$$= \{b_n(M(x_{19}),M(x_{22}),M(x_P));b_n(M(x_{20}),M(x_{23}),M(x_P));b_n(M(x_{21}),M(x_{24}),M(x_P))\}$$

$$(4.4)$$

where,

$$b_n(s,t,p) = \begin{cases} 1 & if(s > p)or(t > p) \\ 1 & if(s < p)or(t < p) \\ 0 & else \end{cases} \qquad (4.5)$$

LQEP is defined by Equations (4.1)–(4.4) as follows.

$$LQEP = \left[ DLE(M(x_P))\big|_{0°}, DLE(M(x_P))\big|_{45°}, DLE(M(x_P))\big|_{90°}, DLE(M(x_P))\big|_{135°} \right]$$

$$(4.6)$$

Subsequently, image under consideration is transformed into patterns possessing values ranging from 0 to 4095. Upon completing computation of LQEP, entire image is depicted by building a histogram based on Equation (4.7).

$$Hist_{LQEP}(b) = \sum_{d=1}^{Q_1}\sum_{e=1}^{Q_2} b_n\left(LQEP(d,e),f\right); f \in [0,4095]; \qquad (4.7)$$

where, LQEP(d,e) shows the LQEP operator value ranging from 0 to 4095.

## 4.2.1 PROPOSED IMAGE RETRIEVAL SYSTEM

In the current work, LQP and DLEP concepts are integrated for content-based image retrieval. Initially, image is loaded and converted into gray scale. Followed by that, four-directional HVDA$_7$ structure is extracted from geometric models. In the next step, maxima or minima (extrema) in four directions, specifically in 0°, 90°, 45° and 135°, are collected. Finally, feature in the form of LQEP is created by building histograms. In order to enhance efficiency of LQEP approach, we combine proposed method with RGB histogram for image retrieval. Information from the latter is complementary to the LQEP method. Framework of LQEP method is provided in Figure 4.3, preceded by Algorithm.

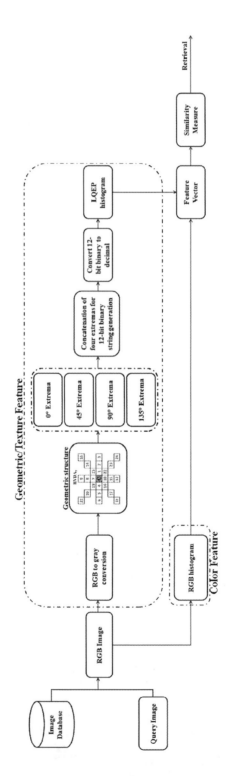

**FIGURE 4.3**  Framework of the proposed LQEP technique.

**ALGORITHM**

*INPUT: DIGITAL IMAGE; OUTPUT: RETRIEVED IMAGE*

1. Load image under test and convert into gray scale.
2. Extract HVDA7 structure for a specified center pixel.
3. Collect extremas in four directions: $0°$, $45°$, $90°$ and $135°$.
4. From directional extrema in four directions, calculate 12-bit LQEP.
5. Build histogram for 12-bit LQEP obtained in step 4.
6. Build RGB histogram from the query image.
7. Create feature vector by concatenating histograms obtained in steps 4 and 5.
8. Match query image and image from database using Equation (1.3).
9. Retrieve images based on similarity level.

## 4.3   EXPERIMENTAL RESULTS AND DISCUSSION

Effectiveness of devised method is examined by performing three experiments on standard repositories. Image databases considered are Corel-1k, Corel-5k and MIT VisTex.

Each image of repository becomes query in all experiments. Proposed method collects $n$ database images $Y = (y_1, y_2, ..., y_m)$ with the shortest matching distance obtained as per Equation 1.3. If retrieved image $y_i = 1, 2, 3 ...., m$ and query image belongs to the same class of database, it can be concluded that system has recognized the predicted image else unsuccessful in identifying the predicted image.

As specified at the beginning, performance of proposed framework is evaluated using performance metrics like average precision/average retrieval precision (ARP) and average recall/average retrieval rate [Equations 1.5–1.10]. As the neighborhood under consideration is higher than the other methods, the ability to extract more spatial information enables this method to achieve good results.

### 4.3.1   Corel-1k Database

In this experiment, Corel-1k database is utilized. We selected 1000 images from 10 different classes with 100 images per domain. Performance of proposed method is evaluated in terms of average retrieval precision and average retrieval rate as specified in Equations 1.5 through 1.10.

From the results provided, it is clear that the proposed method achieves the precision of 76.73, where the recent method LQP achieves only 71.55. Some other popular methods exhibit even much lesser values of precision as compared to the LQEP method.

Table 4.1 and Figure 4.4 depict retrieval results of proposed method and other existing techniques in terms of ARP on Corel–1k database. Table 4.2 and Figure 4.5 illustrate retrieval results of proposed method and other methods in terms of average retrieval recall on Corel–1k database. From Tables 4.1, 4.2, Figures 4.4 and 4.5, it is

**TABLE 4.1**

**Retrieval Results of Proposed LQEP Method and Various Other Existing Methods in Terms of ARP at n = 20 On Corel-1k Database**

| Category | Jhanwar et al. | Lin et al. | cc | Vadivel et al. | S Murala et al. | LBP | LTP | LDP | LTrP | Reddy et al. | LQP | Proposed Method |
|---|---|---|---|---|---|---|---|---|---|---|---|---|
| | | | | | Average Retrieval Precision (ARp) (%); (n = 20) | | | | | | | |
| Africans | 53.15 | 68.3 | 80.4 | 78.25 | 69.75 | 52.46 | 57.2 | 55.35 | 6.9 | 61.3 | 61.95 | 75.15 |
| Beaches | 43.85 | 54 | 41.25 | 44.25 | 54.25 | 51.33 | 43.6 | 52.05 | 53.9 | 51.25 | 53.9 | 57.65 |
| Buildings | 48.7 | 56.2 | 55.65 | 59.1 | 63.95 | 55.66 | 63.35 | 62.25 | 63.4 | 57.85 | 61.95 | 74.7 |
| Buses | 82.8 | 88.8 | 76.7 | 86.5 | 89.65 | 96.33 | 95.5 | 95.8 | 96.55 | 94.4 | 97.4 | 94.3 |
| Dmosatus | 95 | 99.3 | 99 | 98.7 | 98.7 | 95.23 | 96.8 | 94.5 | 98 | 97.85 | 98.75 | 98.95 |
| Elephants | 34.85 | 65.8 | 56.2 | 59 | 48.8 | 42.3 | 46.5 | 43.35 | 46. | 48.9 | 5 | 56.55 |
| Flowers | 88.35 | 89.1 | 92.9 | 85.35 | 92.3 | 85.63 | 91.4 | 85.2 | 86.6 | 89.1 | 91.85 | 95.45 |
| Horses | 59.35 | 8.3 | 76.5 | 74.95 | 89.45 | 65.3 | 64.75 | 69.4 | 72.15 | 66.2 | 77.3 | 86.65 |
| Mountains | 3.8 | 52.2 | 33.7 | 36.55 | 47.3 | 35.93 | 34.55 | 33.55 | 36.1 | 39.4 | 43.75 | 45.9 |
| Food | 5.4 | 73.3 | 7.6 | 64.4 | 7.9 | 7.36 | 7.65 | 76.15 | 75.5 | 75.35 | 78.65 | 82 |
| Total | 58.72 | 72.7 | 68.29 | 68.66 | 72.5 | 65.3 | 66.38 | 66.76 | 68.87 | 68.16 | 71.55 | 76.73 |

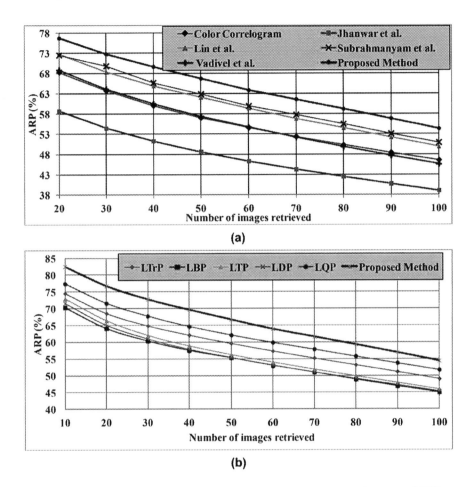

**(a)**

**(b)**

**FIGURE 4.4** Comparison of proposed method with relevant methods in terms of ARP on Corel-1k database.

evident that proposed method shows a considerable improvement as compared to recent methods in terms of precision, ARP, recall and average recall rate on Corel–1k database. Figure 4.6(a) and (b) compares proposed method and similarity distance metrics on Corel-1k database in terms of ARP and average recall rate, respectively. From Figure 4.6, it is noted that $d_1$ distance metric outperforms other distance metrics on Corel-1k database. Figure 4.7 provides query results of proposed approach from Corel-1k database.

### 4.3.2 Corel-5k Database

Corel-5k database is used in this experiment. It contains 5000 images from 50 different categories. Each category has 100 images. Performance of proposed method is determined in terms of ARP and average recall rate.

**TABLE 4.2**

**Retrieval Results of Proposed LQEP Method and Various Other Existing Methods in Terms of ARR at n = 2 On Corel-1K Database**

| Category | Jhanwar et al. | Lin et al. | cc | Vadivel et al. | Subrahmanyam et al. | LBP | LTP | LDP | LTrP | Reddy et al. | LQP | Proposed Method |
|---|---|---|---|---|---|---|---|---|---|---|---|---|
| | | | | | Average Retrieval Recall (ARR) (%) | | | | | | | |
| Africans | 32.21 | 42.1 | 46.29 | 48.41 | 43.58 | 38.1 | 32.9 | 38.1 | 38.6 | 39.25 | 41.23 | 44.38 |
| Beaches | 29.4 | 32.1 | 25.29 | 25.85 | 35.77 | 35.4 | 29.4 | 36.2 | 38.3 | 33.82 | 34.68 | 38.32 |
| Buildings | 27.7 | 36.5 | 35.1 | 37.5 | 34.89 | 33.7 | 35 | 36.5 | 34.9 | 31.96 | 33.55 | 44.84 |
| Buses | 48.66 | 61.7 | 6.97 | 66.52 | 63.39 | 7.5 | 69.9 | 74.2 | 73.4 | 73.57 | 77.85 | 73 |
| Dinosaurs | 81.44 | 94.1 | 89.59 | 78.11 | 92.78 | 75.1 | 87.5 | 77.2 | 83.7 | 9.28 | 92.16 | 92.55 |
| Elephants | 21.42 | 33.1 | 34.14 | 35.66 | 3.31 | 25.4 | 27.8 | 28.5 | 29.5 | 3.53 | 3.91 | 32.78 |
| Flowers | 63.53 | 75 | 77.69 | 57.73 | 64.59 | 65.6 | 71.3 | 62.2 | 65.8 | 69.32 | 76.5 | 79.99 |
| Horses | 35.84 | 47.6 | 36.13 | 41.47 | 66.55 | 42.2 | 4.4 | 44.3 | 43.1 | 36.16 | 47.8 | 5.39 |
| Mountains | 21.75 | 27.7 | 21.2 | 24.37 | 32.9 | 26.9 | 23.6 | 24.6 | 27.5 | 29.35 | 29.55 | 32.32 |
| Food | 29.2 | 49 | 39.27 | 38.24 | 45.12 | 37.2 | 4.5 | 47.9 | 52.2 | 45.3 | 52.25 | 54.54 |
| Total | 39.6 | 49.89 | 46.54 | 45.34 | 5.91 | 44.9 | 45.8 | 46.9 | 48.7 | 47.95 | 51.64 | 54.31 |

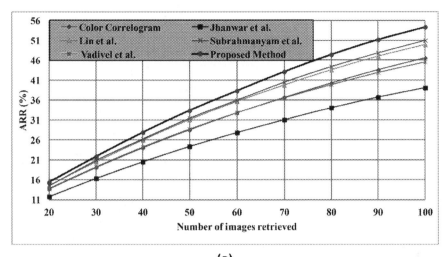

(a)

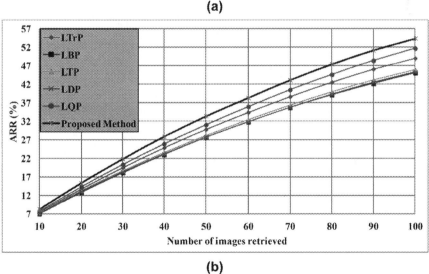

(b)

**FIGURE 4.5 (A) AND (B)**    Comparison of proposed method with relevant methods in terms of ARR on Corel-1k database.

Table 4.3 provides retrieval values of proposed method and other recent methods on Corel-5k database in terms of precision and recall. Figure 4.8(a) and (b) show category-wise performance of methods in terms of ARP and average recall rate on Corel-5k repository. Performance in the form of ARP and average retrieval rate (ARR) on Corel-5k database can be seen in Figure 4.8(c) and (d), respectively. From Table 4.3 and Figure 4.8, it is obvious that devised method shows an improvement as compared to other methods in terms of evaluation measures on Corel-5k database.

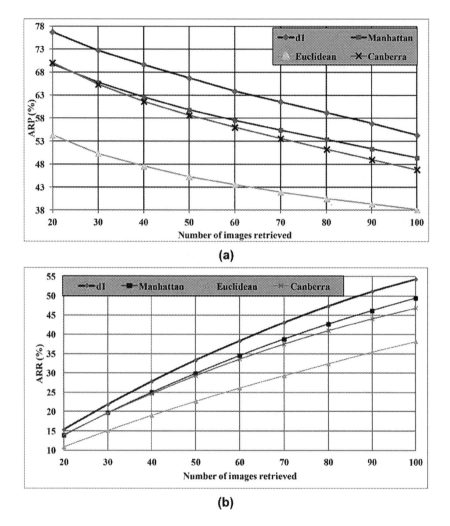

**FIGURE 4.6 (A) AND (B)**    Comparison of proposed method w.r.t distance measures in terms of ARP and ARR on Corel-1k database.

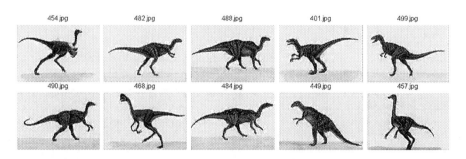

**FIGURE 4.7**    Query results of proposed method on Corel-1k database.

**TABLE 4.3**
**Results of Various Methods in Terms of Precision and Recall On Corel-5k**

| Database | Performance | CS_LBP | LEPSEGR | LEPINVR | BLK_LBPR | LBPR | OLEPR | Reddy et al. | LQPR | PMR |
|---|---|---|---|---|---|---|---|---|---|---|
| | | | | | | Method | | | | |
| Corel-5k | Precision(%) | 32.9 | 41.5 | 35.1 | 45.7 | 43.6 | 48.8 | 54.4 | 5.7 | 62.2 |
| | Recall(%) | 14 | 18.3 | 14.8 | 2.3 | 19.2 | 21.1 | 24.1 | 22.41 | 29.61 |

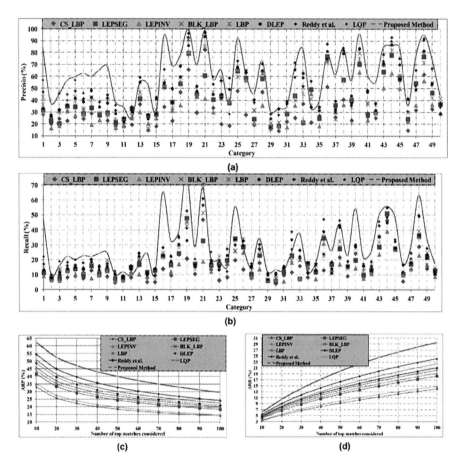

**FIGURE 4.8** Comparison of proposed method and other methods on Corel–5k. (a) Category-wise performance as a measure of precision; (b) Category-wise performance as a measure of recall; (c) Complete database performance as a measure of average retrieval precision; (d) Complete database performance as a measure of average recall rate.

Performance of proposed method is tested with distance measures on Corel-5k database as depicted in Figure 4.9. From Figure 4.9, it is noted that $d_1$ distance outperforms other distance measures in terms of ARP and average recall rate on Corel-5k repository. Figure 4.10 shows query results of proposed method on Corel-5k repository.

### 4.3.3 MIT VisTex Database

MIT VisTex database is considered which contains 40 varieties of textures. 512×512 sized texture is partitioned into 128×128 non-overlapping sub-images, thereby producing a repository of 640 (40×16) images. Each image in database is used as query

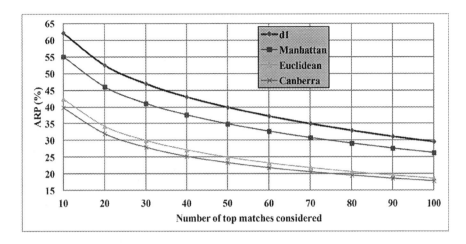

**FIGURE 4.9**  Comparison of different types of distance measures in terms of average retrieval precision on Corel-5k database.

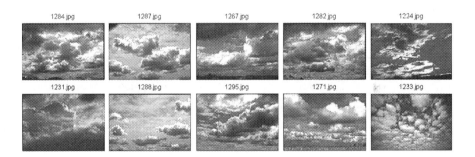

**FIGURE 4.10**  Query results of proposed method on Corel-5k database.

in this experiment. Average retrieval recall or ARR given in Equations 1.5 through 1.10 are used to compare results.

Figures 4.11 and 4.12 depict the performance of different methods in terms of average recall rate and ARP on MIT VisTex database. It is evident that proposed method yields a significant improvement as compared to other approaches as a measure of ARR and ARP on MIT VisTex repository. Figure 4.13 compares performance of proposed method with similarity distance measures in terms of ARR on MIT VisTex database. From Figure 4.13, it is noted that $d_1$ distance measure outperforms other metrics in terms of average recall rate on MIT VisTex database. Figure 4.14 provides query results of proposed method on MIT VisTex database.

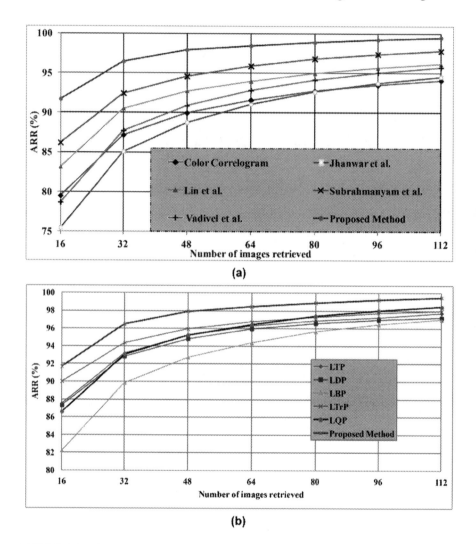

**FIGURE 4.11 (A) AND (B)** Comparison of proposed method with other methods in terms of average retrieval rate.

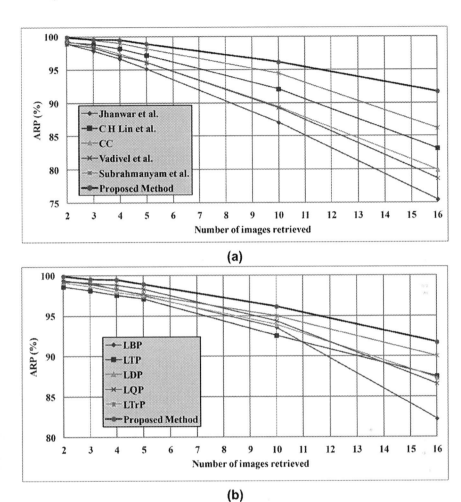

**FIGURE 4.12 (A) AND (B)** Comparison of proposed method with other methods in terms of average retrieval precision.

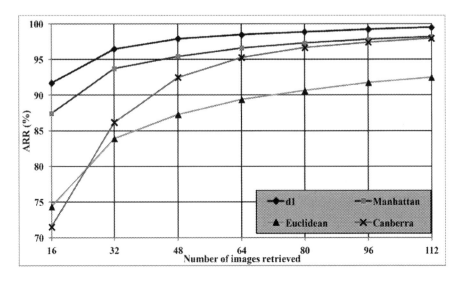

**FIGURE 4.13**   Comparison of proposed method with different distance measures in terms of average recall rate.

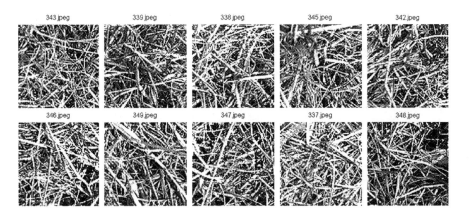

**FIGURE 4.14**   Query results of the proposed method.

## 4.4   CONCLUSION

A new approach for content-based image retrieval is introduced in this chapter. Proposed LQEP differs from LBPs in a way that it explores directional edge information based on local extrema in 0°, 45°, 90° and 135°. Performance of proposed method is tested by experimenting on three standard databases. Retrieval results exhibit a significant improvement in terms of evaluation measures when compared to other approaches.

A novel feature descriptor named local color oppugnant quantized extrema patterns is presented in Chapter 5. It is an extended form of LQEP method targeting the improvement in the performance.

## SOLVED PROBLEMS

1. Extract HVDA$_7$ pattern of the local quantized extrema patterns (LQEP) for the image of 7×7 size.

| 52 | 35 | 26 | 14 | 20 | 15 | 58 |
|----|----|----|----|----|----|----|
| 41 | 46 | 32 | 15 | 64 | 45 | 20 |
| 15 | 23 | 33 | 52 | 41 | 54 | 51 |
| 31 | 25 | 61 | C | 43 | 33 | 51 |
| 28 | 32 | 26 | 38 | 45 | 12 | 54 |
| 30 | 49 | 21 | 7 | 15 | 16 | 17 |
| 15 | 6 | 49 | 19 | 28 | 37 | 42 |

**Ans:** The HVDA$_7$ pattern contains, Horizontal, Vertical, Diagonal and Anti-diagonal patterns, and is shown below.

### H∨DA$_7$ Pattern

|  | 52 |  |  | 14 |  |  | 58 |  |
|---|---|---|---|---|---|---|---|---|
|  |  | 46 |  | 15 |  | 45 |  |  |
|  |  |  | 33 | 52 | 41 |  |  |  |
| 31 | 25 | 61 | C | 43 | 33 | 51 |
|  |  |  | 26 | 38 | 45 |  |  |  |
|  |  | 49 |  | 7 |  | 16 |  |  |
|  | 15 |  |  | 19 |  |  | 42 |  |

2. Extract the 0° and 90° extrema patterns for the image of 5×5 size.

| 42 | 31 | 25 | 12 | 41 |
|----|----|----|----|----|
| 36 | 28 | 19 | 25 | 34 |
| 26 | 34 | C | 31 | 29 |
| 16 | 23 | 17 | 28 | 37 |
| 40 | 38 | 19 | 20 | 25 |

**Ans:** The **0° extrema pattern** is given by

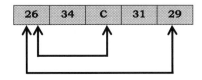

| 26 | 34 | C | 31 | 29 |

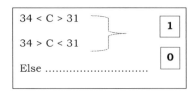

**90° extrema pattern** is given by

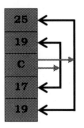

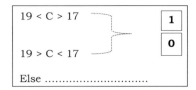

3.  Extract the Disk5 pattern for the image of 7×7 size.

| 15 | 27 | 45 | 21 | 47 | 28 | 9 |
|----|----|----|----|----|----|----|
| 42 | 16 | 31 | 23 | 37 | 8 | 38 |
| 36 | 43 | 17 | 25 | 7 | 30 | 46 |
| 6 | 5 | 4 | C | 1 | 2 | 3 |
| 41 | 10 | 14 | 26 | 18 | 29 | 39 |
| 11 | 13 | 32 | 24 | 40 | 19 | 48 |
| 12 | 35 | 34 | 22 | 33 | 44 | 20 |

**Ans:** The Disk5 pattern for the image of 7×7 size is given as

| 15 | 27 | 45 | 21 | 47 | 28 | 9 |
|----|----|----|----|----|----|----|
| 42 | 16 | 31 | 23 | 37 | 8 | 38 |
| 36 | 43 | 17 | 25 | 7 | 30 | 46 |
| 6 | 5 | 4 | C | 1 | 2 | 3 |
| 41 | 10 | 14 | 26 | 18 | 29 | 39 |
| 11 | 13 | 32 | 24 | 40 | 19 | 48 |
| 12 | 35 | 34 | 22 | 33 | 44 | 20 |

4. Extract the Disk3 pattern for the image of 5×5 size.

| 12 | 17 | 5 | 20 | 13 |
|----|----|---|----|----|
| 21 | 11 | 6 | 14 | 22 |
| 4 | 3 | C | 1 | 2 |
| 24 | 15 | 7 | 10 | 23 |
| 16 | 18 | 8 | 19 | 9 |

**Ans:** The Disk3 pattern is given by

| 12 | 17 | 5 | 20 | 13 |
|----|----|---|----|----|
| 21 | 11 |  | 14 | 22 |
| 4 |  | C |  | 2 |
| 24 | 1 | 7 | 10 | 23 |
| 16 | 18 | 8 | 19 | 9 |

5. Extract the Disk7 pattern for the image of 9×9 size.

| 10 | 15 | 20 | 13 | 5 | 46 | 43 | 91 | 11 |
|----|----|----|----|----|----|----|----|----|
| 24 | 6 | 16 | 14 | 211 | 115 | 88 | 121 | 58 |
| 12 | 3 | 32 | 85 | 1 | 52 | 78 | 27 | 53 |
| 8 | 27 | 54 | 42 | 125 | 93 | 34 | 136 | 19 |
| 23 | 42 | 65 | 36 | c | 79 | 87 | 28 | 24 |
| 13 | 11 | 12 | 38 | 28 | 98 | 56 | 46 | 79 |
| 22 | 22 | 31 | 19 | 52 | 67 | 84 | 61 | 32 |
| 14 | 34 | 57 | 123 | 64 | 36 | 91 | 16 | 81 |
| 15 | 62 | 52 | 78 | 21 | 150 | 101 | 34 | 45 |

**Ans:** The Disk7 pattern is given by

| 10 | 15 | 20 | 13 | 5 | 46 | 43 | 91 | 11 |
|----|----|----|----|----|----|----|----|----|
| 24 | 6 | 16 | 14 | 211 | 115 | 88 | 121 | 58 |
| 12 | 3 | 32 | 85 | 1 | 52 | 78 | 27 | 53 |
| 8 | 27 | 54 | 42 | 125 | 93 | 34 | 136 | 19 |
| 23 | 42 | 65 | 36 | c | 79 | 87 | 28 | 24 |
| 13 | 11 | 12 | 38 | 28 | 98 | 56 | 46 | 79 |
| 22 | 22 | 31 | 19 | 52 | 67 | 84 | 61 | 32 |
| 14 | 34 | 57 | 123 | 64 | 36 | 91 | 16 | 81 |
| 15 | 62 | 52 | 78 | 21 | 150 | 101 | 34 | 45 |

## EXERCISES

1. Extract $HVDA_7$ pattern of the local quantized extrema patterns(LQEP) for the image of 7×7 size.

| 42 | 65 | 81 | 74 | 10 | 26 | 8 |
|----|----|----|----|----|----|----|
| 31 | 58 | 64 | 18 | 4 | 51 | 20 |
| 26 | 95 | 47 | 39 | 11 | 68 | 51 |
| 68 | 31 | 41 | C | 39 | 44 | 22 |
| 38 | 74 | 34 | 71 | 45 | 37 | 54 |
| 40 | 9 | 88 | 54 | 19 | 16 | 66 |
| 21 | 76 | 69 | 19 | 27 | 79 | 66 |

2. Extract the 45° and 135° extrema patterns for the image of 5×5 size.

| 24 | 13 | 52 | 34 | 97 |
|----|----|----|----|----|
| 63 | 78 | 91 | 28 | 56 |
| 62 | 43 | C | 51 | 74 |
| 18 | 32 | 71 | 64 | 62 |
| 4 | 83 | 91 | 26 | 51 |

3. Extract the Disk3 pattern for the image of 5×5 size.

| 21 | 32 | 84 | 30 | 61 |
|----|----|----|----|----|
| 42 | 22 | 45 | 65 | 54 |
| 65 | 46 | C | 74 | 32 |
| 89 | 85 | 14 | 20 | 24 |
| 17 | 9 | 24 | 14 | 7 |

4. Extract the Disk5 pattern for the image of 7×7 size.

| 2 | 6 | 1 | 6 | 0 | 26 | 8 |
|----|----|----|----|----|----|----|
| 1 | 8 | 6 | 8 | 8 | 5 | 2 |
| 6 | 5 | 7 | 9 | 1 | 6 | 5 |
| 8 | 3 | 4 | C | 9 | 4 | 2 |
| 3 | 7 | 3 | 7 | 5 | 7 | 5 |
| 4 | 8 | 8 | 4 | 9 | 1 | 4 |
| 9 | 6 | 9 | 1 | 7 | 9 | 6 |

5. Extract the Disk7 pattern for the image of 11×11 size.

| 21 | 150 | 20 | 13 | 5 | 46 | 43 | 91 | 14 | 52 | 11 |
|----|----|----|----|----|----|----|----|----|----|----|
| 211 | 100 | 16 | 14 | 42 | 115 | 88 | 121 | 54 | 31 | 58 |
| 128 | 98 | 32 | 85 | 64 | 93 | 78 | 27 | 10 | 15 | 53 |
| 8 | 27 | 54 | 42 | 47 | 66 | 34 | 136 | 24 | 6 | 19 |
| 23 | 24 | 125 | 36 | 41 | 79 | 87 | 28 | 12 | 3 | 24 |
| 13 | 36 | 178 | 38 | 34 | C | 56 | 46 | 211 | 12 | 79 |
| 22 | 22 | 31 | 19 | 88 | 67 | 84 | 61 | 17 | 21 | 32 |
| 14 | 34 | 57 | 123 | 69 | 36 | 91 | 16 | 42 | 65 | 81 |
| 15 | 62 | 52 | 78 | 12 | 154 | 101 | 34 | 18 | 12 | 45 |
| 52 | 28 | 60 | 41 | 25 | 4 | 11 | 76 | 64 | 27 | 34 |
| 93 | 47 | 108 | 42 | 47 | 21 | 44 | 66 | 32 | 43 | 55 |

# 5 Local Color Oppugnant Quantized Extrema Patterns

## 5.1 INTRODUCTION

This chapter presents design of a new texture descriptor, local color oppugnant quantized extrema patterns (LCOQEP), which adapts the concept of LOCTP (Jeena Jacob et al. 2014) and local directional extrema patterns (DLEP) (Murala et al. 2012c). Proposed LCOQEP approach extracts the textural structure from two different color planes based on the difference between referenced picture element and neighbors in four directions. Performance of LCOQEP is evaluated on standard image datasets such as Corel-1k, Corel-5k and Corel-10k and ImageNet-25k. It is observed that the proposed method exhibits considerable improvement in precision and recall values compared to other techniques.

## 5.2 LOCAL COLOR OPPUGNANT QUANTIZED EXTREMA PATTERNS

Local operators LOCTP (Jeena Jacob et al. 2014) and DLEP (Murala et al. 2012b) have motivated us to introduce a feature descriptor which is termed as local color oppugnant quantized extrema patterns (LCOQEP). LCOQEP collects the relationship between pixels from two different planes, i.e. relationship between the two planes is collected. For collecting the relationship, we use the red (R), green (G), blue (B) and value (V) planes. Figure 5.1 illustrates the calculation of LCOQEP for a given 7×7 pattern of R, G, B and V color planes. $HVDA_7$ geometric structure is used to extract feature in current work. For a specified direction, detailed description of LCOQEP feature extraction is given below.

From color image $I$, RGB and HSV color channels are considered for the purpose of texture-color feature extraction. Local differences between two oppugnant planes $L_1$, and $L_2$ form a base for the creation of LCOQEP operator as shown in equations below (Equation (5.1), (5.2), (5.3), (5.4)).

$$DE\left(I\left(g_c\right)\right)\Big|_{L_1L_2}^{0°} = \begin{cases} \gamma\left(I\left(g_{1,L_1}\right),I\left(g_{4,L_1}\right),I\left(g_{c,L_2}\right)\right); \\ \gamma\left(I\left(g_{2,L_1}\right),I\left(g_{5,L_1}\right),I\left(g_{c,L_2}\right)\right); \\ \gamma\left(I\left(g_{3,L_1}\right),I\left(g_{6,L_1}\right),I\left(g_{c,L_2}\right)\right) \end{cases} \forall L_1L_2 = RV,GV,BV \quad (5.1)$$

DOI: 10.1201/9781003123514-5

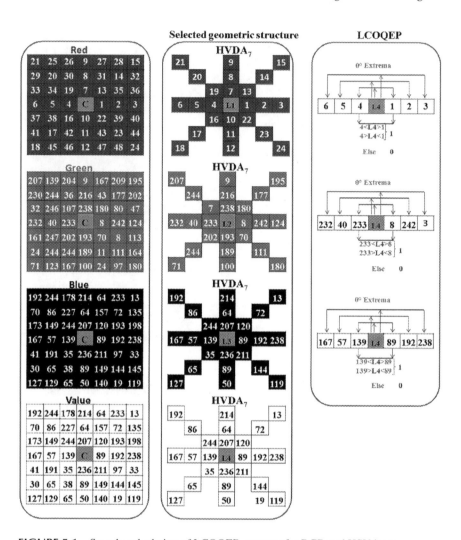

**FIGURE 5.1**  Sample calculation of LCOQEP operator for RGB and HSV image.

$$DE\left(I\left(g_{c}\right)\right)\Big|_{L_{1}L_{2}}^{45^{\circ}} = \begin{cases} \gamma\left(I\left(g_{13,L_{1}}\right),I\left(g_{16,L_{1}}\right),I\left(g_{c,L_{2}}\right)\right); \\ \gamma\left(I\left(g_{14,L_{1}}\right),I\left(g_{17,L_{1}}\right),I\left(g_{c,L_{2}}\right)\right); \\ \gamma\left(I\left(g_{15,L_{1}}\right),I\left(g_{18,L_{1}}\right),I\left(g_{c,L_{2}}\right)\right) \end{cases} \forall L_{1}L_{2} = RV, GV, BV \quad (5.2)$$

$$DE\left(I\left(g_{c}\right)\right)\Big|_{L_{1}L_{2}}^{90^{\circ}} = \begin{cases} \gamma\left(I\left(g_{7,L_{1}}\right),I\left(g_{10,L_{1}}\right),I\left(g_{c,L_{2}}\right)\right); \\ \gamma\left(I\left(g_{8,L_{1}}\right),I\left(g_{11,L_{1}}\right),I\left(g_{c,L_{2}}\right)\right); \\ \gamma\left(I\left(g_{9,L_{1}}\right),I\left(g_{12,L_{1}}\right),I\left(g_{c,L_{2}}\right)\right) \end{cases} \forall L_{1}L_{2} = RV, GV, BV \quad (5.3)$$

$$DE\left(I\left(g_c\right)\right)\Big|_{L_1L_2}^{135°} = \begin{cases} \gamma\left(I\left(g_{19,L_1}\right),I\left(g_{22,L_1}\right),I\left(g_{c,L_2}\right)\right); \\ \gamma\left(I\left(g_{20,L_1}\right),I\left(g_{23,L_1}\right),I\left(g_{c,L_2}\right)\right); \\ \gamma\left(I\left(g_{21,L_1}\right),I\left(g_{24,L_1}\right),I\left(g_{c,L_2}\right)\right) \end{cases} \forall L_1L_2 = RV,GV,BV \quad (5.4)$$

where,

$$\gamma\left(b_1,b_2,d\right) = \begin{cases} 1 & if\left(b_1 > d\right)or\left(b_2 > d\right) \\ 1 & if\left(b_1 < d\right)or\left(b_2 < d\right) \\ 0 & otherwise \end{cases} \quad (5.5)$$

LCOQEP is determined by using Equation (5.6) as follows:

$$LCOQEP_{L_1L_2} = \left[ DE\left(I\left(g_c\right)\right)\Big|_{L_1L_2}^{0°}, DE\left(I\left(g_c\right)\right)\Big|_{L_1L_2}^{45°}, DE\left(I\left(g_c\right)\right)\Big|_{L_1L_2}^{90°}, DE\left(I\left(g_c\right)\right)\Big|_{L_1L_2}^{135°} \right]$$
$$\forall L_1L_2 = RV,GV,BV$$

$$(5.6)$$

Subsequently, given image is transformed into three LCOQEP plots consisting of the values lying between 0 and 4095. After calculating LCOQEP, complete image is represented by constructing a histogram based on Equation (5.7).

$$Hist_{LCOQEP_{L_1L_2}}(l) = \sum_{u=1}^{B_1}\sum_{v=1}^{B_2}\Omega\left(LCOQEP_{L_1L_2}\left(u,v\right),i\right); i \in \left[0,4095\right];$$
$$\forall L_1L_2 = RV,GV,BV$$

$$(5.7)$$

## 5.2.1 PROPOSED IMAGE RETRIEVAL SYSTEM

In the proposed work, we integrate local quantized extrema calculation and oppugnant planes relationship. Oppugnant planes which are considered for quantized extrema calculation are $RV$, $GV$ and $BV$. Structure $L_1L_2$ represents quantized extrema calculation between the pixels of $L_1$ and $L_2$. Here, $L_2$ is considered as the center pixel and $L_1$ is considered as neighbors. Figure 5.2 illustrates the schematic diagram of proposed image indexing and retrieval system preceded by the algorithm. Figure 5.3 illustrates the feature maps that are brought out using the proposed feature extraction approach.

**ALGORITHM**

*i/p: Digital image; o/p: Retrieval outcome*

1. Load color image (RGB) and transform into HSV form.
2. For a given center pixel, capture the neighbors for R, G, B and V planes.
3. Collect required bits based from RV, GV, BV structures.
4. Compute the quantized extrema pattern bits for RV, GV, BV.
5. Extract the LCOQEPs for RV, GV, BV.
6. Build the histograms for RV, GV, BV.
7. Create the histograms for H and S planes of color image.
8. Create feature vector by adjoining histograms.
9. Correlate the query image and database images.
10. Fetch images based on the level of similarity.

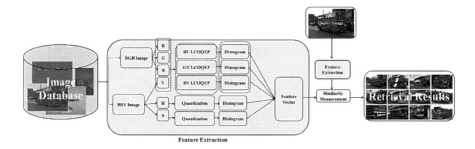

**FIGURE 5.2**    Framework of proposed LCOQEP.

**FIGURE 5.3**    LCOQEP feature maps extracted from a sample image.

The algorithm shown above explains the process of extracting the color information from two different planes. This is an enhancement to LQEP method, which is limited to gray-scale images. The motive behind this algorithm is to improve the precision and recall values of the retrieval systems.

## 5.3  EXPERIMENTAL RESULTS AND DISCUSSION

Capability of proposed technique is determined by performing experiments on standard image repositories. Databases such as Corel-1k, Core-5k, Corel-10k and ImageNet-25k are considered in the process.

During entire stretch of experimentation, every image of repository is considered as the query image. For every query, proposed method explores m images $Z = (z_1, z, ..., z_m)$ according to the distance calculation. If the obtained image $z_i = 1, 2, ...., m$ belongs to the same category of the input image, then we conclude that the framework correctly recognized the predicted outcome, else it is unsuccessful in finding a desired image.

Performance of LCOQEP approach is determined as a measure of average precision/ average retrieval precision, average recall/average retrieval rate as mentioned in equations 1.5–1.10 in Chapter 1. It is apparent from the experiments that the extraction of color information improves the performance.

### 5.3.1  COREL-1K DATABASE

In the first phase of experimentation, Corel-1k image database [38] is used. Performance of various methods as a measure of average retrieval precision on Corel-1k database is provided in Table 5.1 and Figure 5.4. Table 5.2 and Figure 5.5 interpret retrieval output of LCOQEP method and other existing approaches as a measure of ARR on Corel-1k database. From Table 5.1, Table 5.2, Figure 5.4 and Figure 5.5, it is obvious that proposed technique exhibits a considerable increase when compared to advanced methods with respect to average retrieval precision average retrieval rate, tested with Corel-1k dataset. Figure 5.6(a) and (b) shows performance of devised LCOQEP approach with different types of similarity metrics on Corel-1k repository in the forms of average recall rate and average retrieval precision, respectively. From Figure 5.6, it can be noticed that $d_1$ distance calculation method supersedes other distance calculation methods in the forms of ARR and the ARP on Corel-1k database. Figure 5.7 interprets query results on Corel-1k repository.

### 5.3.2  COREL-5K DATABASE

Corel-5k database is used in this experiment. It consists of 5000 images collected from 50 different categories. Each category has 100 images. Performance of proposed method is determined in terms of average retrieval precision and average recall rate. Table 5.3 depicts image retrieval results on Corel-5k as a measure of average retrieval precision and average recall rate. Category-wise performance of various methods in terms of average retrieval precision and average recall rate is shown in

**TABLE 5.1**

**Retrieval Results of Proposed LCOQEP Method and Various Other Existing Methods in Terms of ARP at n=20 on Corel-1k Database**

| Category | Jhanwar et al. | Lin et al. | cc | Vadivel et al. | S Murala et al. | LBP | LTP | LDP | LTrP | Reddy et al. | LOCSEP | LCOQEP |
|---|---|---|---|---|---|---|---|---|---|---|---|---|
| | | | | | Average Precision (%); (n=20) | | | | | | | |
| Africans | 53.1 | 68.3 | 80.4 | 78.2 | 69.7 | 52.4 | 57.2 | 55.3 | 60.9 | 61.3 | 73.6 | 81.6 |
| Beaches | 43.85 | 54 | 41.2 | 44.2 | 54.2 | 51.3 | 43.6 | 52.05 | 53.9 | 51.2 | 46.3 | 56.3 |
| Buildings | 48.7 | 56.2 | 55.6 | 59.1 | 63.9 | 55.6 | 63.35 | 62.2 | 63.4 | 57.8 | 74.3 | 80.2 |
| Buses | 82.8 | 88.8 | 76.7 | 86 | 89.6 | 96.3 | 95.5 | 95.8 | 96.55 | 94.4 | 92.2 | 97 |
| Dinosaurs | 95 | 99.3 | 99 | 98.7 | 98.7 | 95.2 | 96.8 | 94.5 | 98 | 97.8 | 99.6 | 99.3 |
| Elephants | 34.8 | 65.8 | 56.2 | 59 | 48.8 | 42 | 46 | 43.3 | 46.1 | 48.9 | 54.1 | 72.2 |
| Flowers | 88.3 | 89.1 | 92.9 | 85.3 | 92.3 | 85.6 | 91.4 | 85.2 | 86.6 | 89.1 | 92 | 96.6 |
| Horses | 59.3 | 80.3 | 76.5 | 74.9 | 89.4 | 65.3 | 64.75 | 69.4 | 72.15 | 66.2 | 94.3 | 97.2 |
| Mountains | 30.8 | 52.2 | 33.7 | 36.5 | 47.3 | 35.9 | 34.55 | 33.55 | 36.1 | 39.4 | 50 | 61.2 |
| Food | 50.4 | 73.3 | 70.6 | 64.4 | 70.9 | 70.3 | 70.65 | 76.1 | 75.05 | 75.3 | 82.7 | 90.7 |
| Total | 58.7 | 72.7 | 68.2 | 68.6 | 72.5 | 65 | 66.38 | 66.7 | 68.87 | 68.1 | 75.9 | 78.4 |

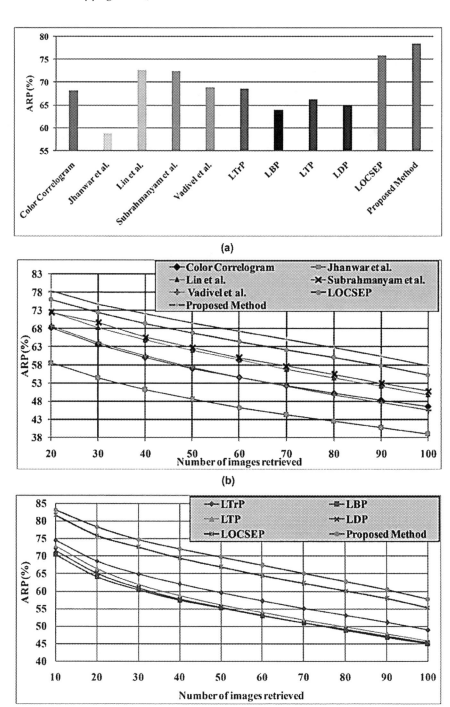

**FIGURE 5.4** Comparison of related methods in terms of average retrieval precision on Corel-1k image database.

**TABLE 5.2**

**Retrieval Results of Proposed LCOQEP Method and Various Other Existing Methods in Terms of ARR at n=20 on Corel-1k Database**

| Category | Average recall (%);(n=20) | | | | | | | | | | | |
|---|---|---|---|---|---|---|---|---|---|---|---|---|
| | Jhanwar et al. | Lin et al | cc | Vadivel et al. | S Murala | LBP | LTP | LDP | LTrP | Reddy et al | LOCSEP | LCOQEP |
| Africans | 32.21 | 42.1 | 46.3 | 48.41 | 43.58 | 38.1 | 32.9 | 38.1 | 38.6 | 39.25 | 47.26 | 50.44 |
| Beaches | 29.04 | 32.1 | 25.3 | 25.85 | 35.77 | 35.4 | 29.4 | 36.2 | 38.3 | 33.82 | 29.07 | 30.66 |
| Buildings | 27.7 | 36.5 | 35 | 37.05 | 34.89 | 33.7 | 35 | 36.5 | 34.9 | 31.96 | 44.43 | 41.52 |
| Buses | 48.66 | 61.7 | 61 | 66.52 | 63.39 | 70.5 | 69.9 | 74.2 | 73.4 | 73.57 | 67.48 | 71.82 |
| Dinosaurs | 81.44 | 94.1 | 89.6 | 78.11 | 92.78 | 75.1 | 87.5 | 77.2 | 83.7 | 90.28 | 94.07 | 92.6 |
| Elephants | 21.42 | 33.1 | 34.1 | 35.66 | 30.31 | 25.4 | 27.8 | 28.5 | 29.5 | 30.53 | 31.19 | 34.43 |
| Flowers | 63.53 | 75 | 77.7 | 57.73 | 64.59 | 65.6 | 71.3 | 62.2 | 65.8 | 69.32 | 72.11 | 79.97 |
| Horses | 35.84 | 47.6 | 36.1 | 41.47 | 66.55 | 42.2 | 40.4 | 44.3 | 43.1 | 36.16 | 77.64 | 79.22 |
| Mountains | 21.75 | 27.7 | 21 | 24.37 | 32.09 | 26.9 | 23.6 | 24.6 | 27.5 | 29.35 | 33.59 | 37.46 |
| Food | 29.02 | 49 | 39.3 | 38.24 | 45.12 | 37.2 | 40.5 | 47.9 | 52.2 | 45.3 | 56.38 | 59.97 |
| Total | 39.06 | 49.9 | 46.5 | 45.34 | 50.91 | 44.9 | 45.8 | 46.9 | 48.7 | 47.95 | 55.32 | 57.8 |

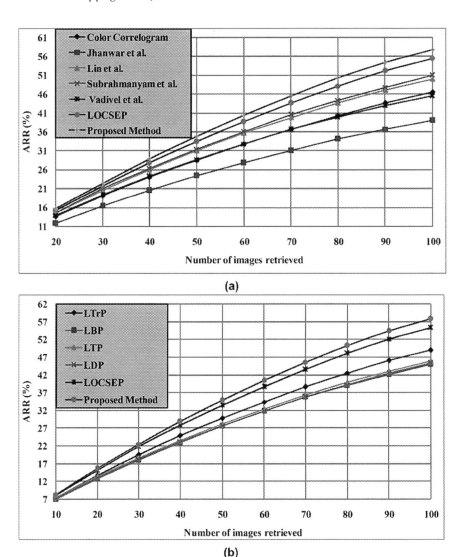

**FIGURE 5.5**    Comparison of related methods as a measure of average recall rate on Corel-1k image repository.

Figure 5.8(a) and Figure 5.8(b), respectively. Further, performance of different techniques is also evaluated against average recall rate and average retrieval precision on Corel-5k database. The ARR and ARP results are given in Figure 5.9(a) and Figure 5.9(b), respectively.

From Table 5.3, Figure 5.8 and Figure 5.9, it is evident that proposed method shows a significant increase as compared to the advanced frameworks on Corel-5k repository. Figure 5.10 depicts query results of LCOQEP.

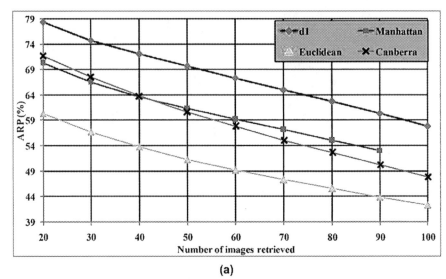

(a)

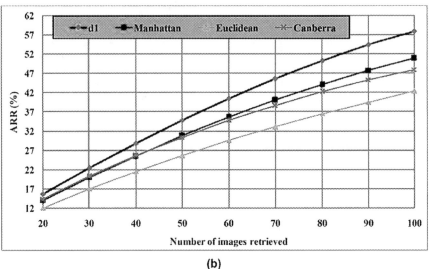

(b)

**FIGURE 5.6** Comparison of related methods for distance measures of average retrieval precision and average recall rate.

### 5.3.3 COREL-10K DATABASE

Corel-10k database is utilized in this experiment. Corel-10k repository is a composition of 10,000 images from 100 different categories with 100 images per domain. Performance evaluation metrics considered here are average retrieval precision and average recall rate.

**FIGURE 5.7**    Results of retrieval from Corel-1k database using proposed method.

Domain-wise performance of various methods in terms of precision and recall are presented in Figure 5.11(a) and (b), respectively. Further, performance is also obtained in terms of average retrieval precision and average retrieval rate. Results of the same are provided in Figure 5.12(a) and (b) respectively. From Table 5.3, Figure 5.11 and Figure 5.12, it is obvious that introduced approach shows a considerable increase in performance as compared to other advanced methods on Corel-10k repository.

### 5.3.4 IMAGENET-25K DATABASE

In this experiment, ImageNet is considered for image retrieval. It contains 25,000 images from different categories. Performance of the proposed method is measured in terms of average retrieval precision. Figure 5.13 depicts the performance of various methods in terms of average retrieval precision on ImageNet-25k database. From the Figure 5.13, it is clear that the devised method outperforms other popular methods. The improvement is due to the extraction of color information along with the texture.

**TABLE 5.3**
**Results of Various Methods in Terms of Precision and Recall on Corel-5k and Corel-10k Databases**

| Database | Performance | Method | | | | | | | | |
|---|---|---|---|---|---|---|---|---|---|---|
| | | CS_LBP | LEPSEG | LEPINV | BLK_LBP | LBP | DLEP | MDLEP | LOCSEP | LCOQEP |
| Corel-5K | Precision (%) | 32.9 | 41.5 | 35.1 | 45.7 | 43.6 | 48.8 | 54.4 | 57.3 | 59.2 |
| | Recall (%) | 14 | 18.3 | 14.8 | 20.3 | 19.2 | 21.1 | 24.1 | 26.9 | 27.4 |
| Corel-10K | Precision (%) | 26.4 | 34 | 28.9 | 38.1 | 37.6 | 40 | 45.4 | 46.8 | 49.4 |
| | Recall (%) | 10.1 | 13.8 | 11.2 | 15.3 | 14.9 | 15.7 | 18.4 | 19.8 | 20.9 |

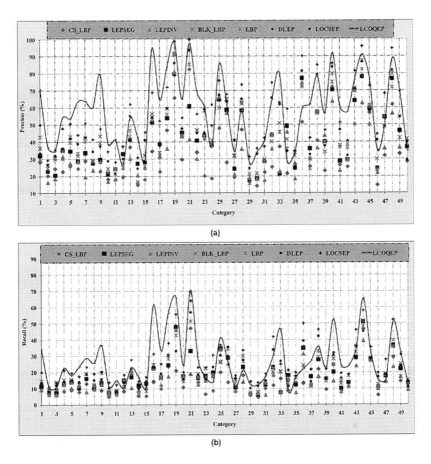

**FIGURE 5.8** Comparison of related methods with proposed LCOQEP method: (a) class-wise precision and (b) class-wise recall.

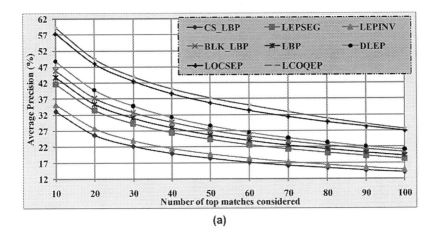

**FIGURE 5.9** ARR and ARP comparison of various methods on Corel-5k database.

(*Continued*)

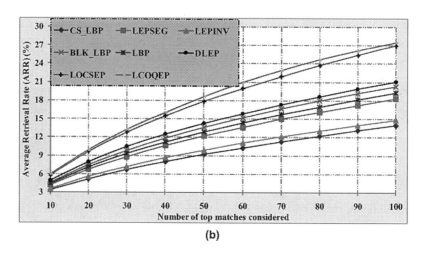

(b)

**FIGURE 5.9 (Continued)**   ARR and ARP comparison of various methods on Corel-5k database.

**FIGURE 5.10**   Retrieved images from Corel-5k repository.

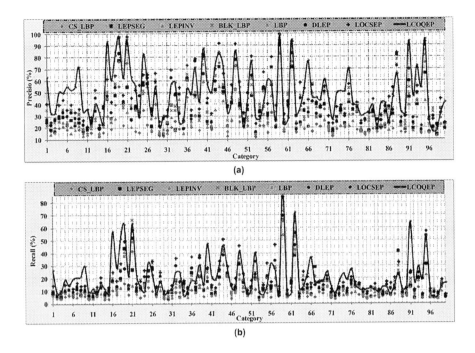

**FIGURE 5.11** Comparison of LCOQEP and other methods on Corel–10k. (a) Category-wise precision, (b) Category-wise recall.

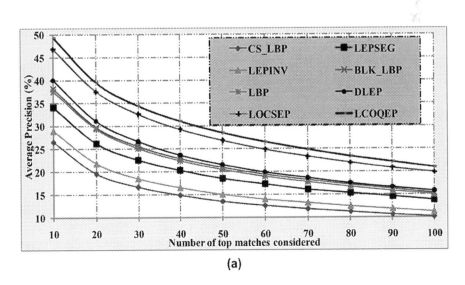

**FIGURE 5.12** Comparison of different methods as a measure of (a) Average Precision and (b) Average retrieval rate. *(Continued)*

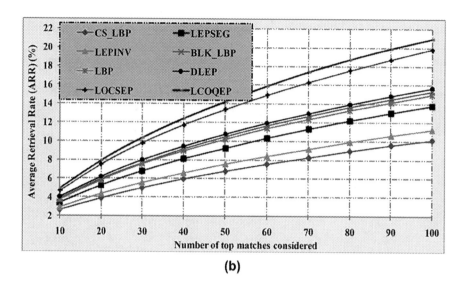

**(b)**

**FIGURE 5.12 (Continued)**    Comparison of different methods as a measure of (a) Average Precision and (b) Average retrieval rate.

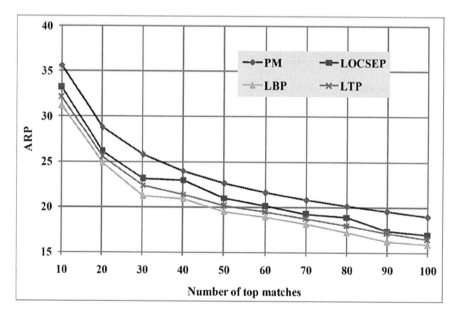

**FIGURE 5.13**   Comparison of various methods as a measure of ARP on ImageNet-25k database. PM: Proposed Method.

## 5.4   CONCLUSION

A new approach for content-based image retrieval is presented in this chapter. LCOQEP extracts directional edge information based on local quantized extrema in $0°, 45°, 90°$ and $135°$ directions from different color planes. Performance of LCOQEP method is observed by experimenting with various databases. Significant improvement is recorded in the accuracy measured in terms of precision and recall.

By considering the pixels at alternate positions, a novel feature descriptor called local mesh quantized extrema patterns is presented in Chapter 6.

## SOLVED PROBLEMS

1. Illustrate the LCOQEP for RGB image of 7×7 size.

   Ans: Three components of RGB image were shown for LCOQEP pattern of HVDA$_7$ pattern.

| 14 | 48 | 2 | 67 | 28 |
|----|----|---|----|----|
| 95 | 15 | 1 | 27 | 110 |
| 79 | 107 | P | 80 | 51 |
| 120 | 26 | 6 | 16 | 99 |
| 25 | 49 | 7 | 68 | 17 |

| 9 | 11 | 1 | 14 | 15 |
|---|----|---|----|----|
| 20 | 17 | 2 | 10 | 16 |
| 6 | 5 | P | 8 | 7 |
| 30 | 25 | 3 | 13 | 18 |
| 29 | 26 | 4 | 33 | 19 |

| 1 | 17 | 11 | 20 | 8 |
|---|----|----|----|---|
| 13 | 2 | 21 | 7 | 14 |
| 10 | 18 | P | 24 | 9 |
| 15 | 6 | 23 | 3 | 16 |
| 5 | 22 | 12 | 19 | 4 |

2. Extract the $0°$ extrema pattern of R component for LCOQEP of RGB image of 5×5 size shown below

| 14 | 48 | 2 | 67 | 28 |
|----|----|---|----|----|
| 95 | 15 | 1 | 27 | 110 |
| 79 | 107 | P | 80 | 51 |
| 120 | 26 | 6 | 16 | 99 |
| 25 | 49 | 7 | 68 | 17 |

Ans: $0°$ extrema of R component in an RGB image is

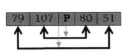

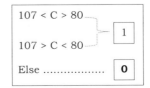

107 < C > 80
107 > C < 80 ........ 1

Else ................. 0

3. Extract the 90° extrema pattern of G component for LCOQEP of RGB image of 5×5 size shown below

| 9 | 11 | 1 | 14 | 15 |
|---|---|---|---|---|
| 20 | 17 | 2 | 10 | 16 |
| 6 | 5 | **P** | 8 | 7 |
| 30 | 25 | 3 | 13 | 18 |
| 29 | 26 | 4 | 33 | 19 |

Ans: 90° extrema pattern is given by

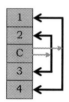
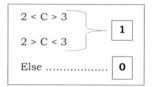

4. Extract the 90° extrema pattern of B component for LCOQEP of RGB image of 5×5 size shown below

| 1 | 17 | 11 | 20 | 8 |
|---|---|---|---|---|
| 13 | 2 | 21 | 7 | 14 |
| 10 | 18 | P | 24 | 9 |
| 15 | 6 | 23 | 3 | 16 |
| 5 | 22 | 12 | 19 | 4 |

Ans: 90° extrema pattern is given by

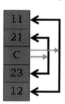
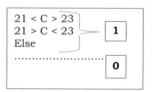

5. Illustrate the H, V, $HV_{11}$, $HVDA_{11}$ structures for the HSV image of 11×11 size.

11×11 HSV image:

| 11 | 40 | 77 | 93 | 63 | 5 | 61 | 83 | 121 | 43 | 31 |
|---|---|---|---|---|---|---|---|---|---|---|
| 88 | 12 | 94 | 38 | 118 | 4 | 90 | 39 | 89 | 30 | 117 |
| 47 | 129 | 13 | 78 | 62 | 3 | 52 | 111 | 29 | 72 | 44 |
| 87 | 55 | 64 | 14 | 48 | 2 | 67 | 28 | 84 | 65 | 109 |
| 32 | 75 | 34 | 95 | 15 | 1 | 27 | 110 | 35 | 106 | 33 |
| 112 | 76 | 58 | 79 | 107 | P | 80 | 51 | 60 | 66 | 100 |
| 54 | 59 | 53 | 120 | 26 | 6 | 16 | 99 | 96 | 50 | 97 |
| 73 | 135 | 72 | 25 | 49 | 7 | 68 | 17 | 142 | 70 | 101 |
| 46 | 74 | 24 | 122 | 81 | 8 | 69 | 71 | 18 | 143 | 45 |
| 86 | 23 | 82 | 37 | 92 | 9 | 85 | 36 | 103 | 19 | 98 |
| 21 | 41 | 115 | 57 | 131 | 10 | 141 | 91 | 56 | 42 | 20 |

Ans:

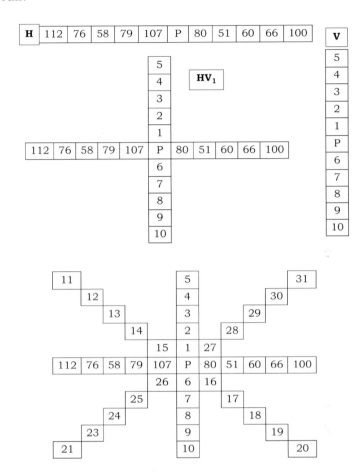

## EXERCISES

1. Illustrate the H, V, $HV_9$, $HVDA_9$ structures for the HSV image of 9×9 size.

   9×9 HSV image:

| 12 | 94 | 38 | 118 | 4 | 90 | 39 | 89 | 30 |
|---|---|---|---|---|---|---|---|---|
| 129 | 13 | 78 | 62 | 3 | 52 | 111 | 29 | 72 |
| 55 | 64 | 14 | 48 | 2 | 67 | 28 | 84 | 65 |
| 75 | 34 | 95 | 15 | 1 | 27 | 110 | 35 | 106 |
| 76 | 58 | 79 | 107 | P | 80 | 51 | 60 | 66 |
| 59 | 53 | 120 | 26 | 6 | 16 | 99 | 96 | 50 |
| 135 | 72 | 25 | 49 | 7 | 68 | 17 | 142 | 70 |
| 74 | 24 | 122 | 81 | 8 | 69 | 71 | 18 | 143 |
| 23 | 82 | 37 | 92 | 9 | 85 | 36 | 103 | 19 |

2.  Illustrate the H, V, $HV_7$, $HVDA_7$ structures for the HSV image of 7×7 size.

    7×7 HSV image:

    | 13 | 78 | 62 | 3 | 52 | 111 | 29 |
    |----|-----|-----|---|----|-----|-----|
    | 64 | 14 | 48 | 2 | 67 | 28 | 84 |
    | 34 | 95 | 15 | 1 | 27 | 110 | 35 |
    | 58 | 79 | 107 | P | 80 | 51 | 60 |
    | 53 | 120 | 26 | 6 | 16 | 99 | 96 |
    | 72 | 25 | 49 | 7 | 68 | 17 | 142 |
    | 24 | 122 | 81 | 8 | 69 | 71 | 18 |

3.  Illustrate the H, V, $HV_5$, $HVDA_5$ structures for the HSV image of 5×5 size.

    | 14 | 48 | 2 | 67 | 28 |
    |-----|-----|---|----|-----|
    | 95 | 15 | 1 | 27 | 110 |
    | 79 | 107 | P | 80 | 51 |
    | 120 | 26 | 6 | 16 | 99 |
    | 25 | 49 | 7 | 68 | 17 |

4.  Extract the $90^0$ extrema pattern of R component for LCOQEP of RGB image of 5×5 size shown below

    | 14 | 48 | 2 | 67 | 28 |
    |-----|-----|---|----|-----|
    | 95 | 15 | 6 | 27 | 110 |
    | 79 | 107 | P | 80 | 51 |
    | 120 | 26 | 1 | 16 | 99 |
    | 25 | 49 | 7 | 68 | 17 |

5.  Extract the $0^0$ extrema pattern of B component for LCOQEP of RGB image of 5×5 size shown below

    | 1 | 17 | 11 | 20 | 8 |
    |----|-----|-----|-----|-----|
    | 13 | 2 | 21 | 7 | 14 |
    | 10 | 18 | P | 24 | 9 |
    | 15 | 6 | 23 | 3 | 16 |
    | 5 | 22 | 12 | 19 | 4 |

6.  Extract the $0^0$ extrema pattern of G component for LCOQEP of RGB image of 5×5 size shown below

    | 9 | 11 | 1 | 14 | 15 |
    |----|-----|---|----|-----|
    | 20 | 17 | 2 | 10 | 16 |
    | 6 | 5 | P | 8 | 7 |
    | 30 | 25 | 3 | 13 | 18 |
    | 29 | 26 | 4 | 33 | 19 |

7. Draw the geometric structures for LCOQEP of G component of RGB image of 5×5 size.

| 15 | 18 | 7 | 16 | 19 |
|----|----|----|----|----|
| 20 | 17 | 2 | 10 | 16 |
| 60 | 45 | P | 8 | 72 |
| 30 | 25 | 9 | 13 | 18 |
| 29 | 26 | 12 | 33 | 19 |

8. Extract geometric extrema pattern of R component for LCOQEP of RGB image of 5×5 size shown below

| 14 | 48 | 2 | 67 | 28 |
|----|----|----|----|----|
| 95 | 15 | 6 | 27 | 110 |
| 79 | 107 | P | 80 | 51 |
| 120 | 26 | 1 | 16 | 99 |
| 25 | 49 | 7 | 68 | 17 |

9. Extract the geometric extrema pattern of B component for LCOQEP of RGB image of 5×5 image shown below

| 1 | 17 | 11 | 20 | 8 |
|----|----|----|----|----|
| 13 | 2 | 21 | 7 | 14 |
| 10 | 18 | P | 24 | 9 |
| 15 | 6 | 23 | 3 | 16 |
| 5 | 22 | 12 | 19 | 4 |

# 6 Local Mesh Quantized Extrema Patterns

## 6.1 INTRODUCTION

Local quantized patterns (Hussain and Triggs 2012) and directional local extrema patterns (Murala et al. 2012a) motivated us to design local mesh quantized extrema patterns (LMeQEP) for image retrieval. Major contributions of the work are summarized as follows: (i) a mesh quantized HVDA structure is explored from image. (ii) Local extrema is extracted from mesh structures to produce LMeQEP. (iii) To achieve improved performance, LMeQEP and RGB histogram are integrated. (iv) Experimentation is done on benchmark image databases.

## 6.2 LOCAL MESH QUANTIZED EXTREMA PATTERNS

Information pertaining to spatial structure of texture is extracted by applying mesh concept on local quantized extrema patterns. More discerning information can be obtained by structuring a mesh using the pixels at alternate positions without scarifying the connectivity of pixel information. Calculation of Local Mesh Quantized Extrema Patterns (LMeQEP) is provided in Figure 6.1.

An approach to collect geometric $HVDA_5$ structure for a pixel (P) at the center in an image I is shown in Figure 6.1. Maxima or minima in four directions, i.e. $0°$, $90°$, $45°$ and $135°$ are computed as follows:

$$DME(I(p_c))\big|_{0°} =$$
$$\left\{ g_2(I(p_{24}), I(p_{23}), I(p_C)); g_2(I(p_{23}), I(p_C), I(p_{21})); g_2(I(p_C), I(p_{21}), I(p_{22})) \right\} \quad (6.1)$$

$$DME(I(p_c))\big|_{45°} =$$
$$\left\{ g_2(I(p_{32}, I(p_{31}), I(p_C)); g_2(I(p_{31}), I(p_C), I(p_{29})); g_2(I(p_C), I(p_{29}), I(p_{30})) \right\} \quad (6.2)$$

$$DME(I(p_c))\big|_{90°} =$$
$$\left\{ g_2(I(p_{26}), I(p_{25}), I(p_C)); g_2(I(p_{25}), I(p_C), I(p_{27})); g_2(I(p_C), I(p_{27}), I(p_{28})) \right\} \quad (6.3)$$

$$DME(I(p_c))\big|_{135°} =$$
$$\left\{ g_2(I(p_{34}), I(p_{33}), I(p_C)); g_2(I(p_{33}), I(p_C), I(p_{35})); g_2(I(p_C), I(p_{35}), I(p_{36})) \right\} \quad (6.4)$$

DOI: 10.1201/9781003123514-6

| 36 | 37 | 26 | 38 | 30 |
|----|----|----|----|----|
| 39 | 35 | 25 | 29 | 40 |
| 24 | 23 | P | 21 | 22 |
| 41 | 31 | 27 | 33 | 42 |
| 32 | 43 | 28 | 44 | 34 |

Pattern

| 36 |    | 26 |    | 30 |
|----|----|----|----|----|
|    | 35 | 25 | 29 |    |
| 24 | 23 | P | 21 | 22 |
|    | 31 | 27 | 33 |    |
| 22 |    | 28 |    | 26 |

Selected geometric structure

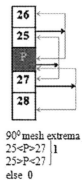

90° mesh extrema
25<P>27 }1
25>P<27 }
else 0

**FIGURE 6.1** LMeQEP calculation using $HVDA_5$ geometric structure.

where

$$g_2(p,q,r) = \begin{cases} 1 & if\,(p > q)\,or\,(r > q) \\ 1 & if\,(p < q)\,or\,(r < q) \\ 0 & otherwise \end{cases} \tag{6.5}$$

LMeQEP is defined according to Equations (6.1), (6.2), (6.3) and (6.4) as follows:

$$LMeQEP = \left[ DME(I(p_c))\big|_{0°}, DME(I(p_c))\big|_{45°},\ DME(I(p_c))\big|_{90°}, DME(I(p_c))\big|_{135°} \right] \tag{6.6}$$

Subsequently, image under test is transformed to LMeQEP maps possessing the patterns ranging from 0 to 4095. Histogram is constructed to represent LMeQEP as specified in Equation 6.7.

$$Hist_{LMeQEP}(p) = \sum_{m=1}^{B_1} \sum_{n=1}^{B_2} g_2(LMeQEP(m,n),p);\, p \in [0,4095]; \tag{6.7}$$

### 6.2.1 PROPOSED IMAGE RETRIEVAL SYSTEM

In this chapter, we present a new image feature descriptor LMeQEP [159] for retrieval by structuring a mesh on the extracted geometric structure. Initially, image is loaded and converted into gray scale. $HVDA_5$ structure representing four directions is collected using the local quantized patterns method. Extremas in 0°, 90°, 45° and 135° are calculated on mesh structure as given in Equations (6.1), (6.2)–(6.4). Subsequently, LMeQEP-based feature descriptor is generated by constructing histogram. To make this method more effective for retrieval, RGB histogram is added to LMeQEP. In this case, color information is complementary to the texture which is depicted in the form of LQEP.

Algorithm for the LMeQEP is given below, followed by framework of the devised method (Figure 6.2). The motive behind the development of this algorithm is to enhance the performance of the system. It is totally a new way of comparing the pixels at different positions, thus creating a mesh structure.

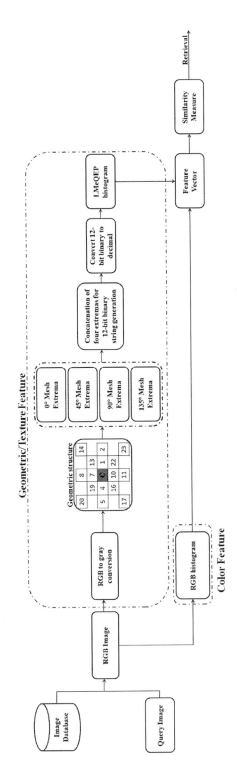

**FIGURE 6.2**  Framework of the proposed LMeEP.

**ALGORITHM**

1. Convert RGB image into gray-scale image.
2. Derive $HVDA_5$ structure for a pixel at the center.
3. Calculate local maxima or minima in $0°$, $90°$, $45°$ and $135°$.
4. Form a 12-bit LMeQEP using four-directional extrema.
5. Build a histogram for 12-bit LMeQEP.
6. Prepare RGB histogram from RGB image.
7. Create a feature descriptor by concatenating RGB and LMeQEP histograms.
8. Compare query and database images.
9. Retrieve images according to similarity index.

## 6.3   EXPERIMENTAL RESULTS AND DISCUSSION

Experiments are conducted on standard datasets such as Corel-1k, and MIT VisTex to determine performance of LMeQEP approach.

Every image in the database becomes query image during experimentation. System returns "n" repository images M= $(m_1, m_2...m_n)$ for each query according to the distance calculated using equation (17). If the output image $m_i$=1, 2...$n$ belongs to the group of query image, it can be stated that retrieval is correctly done, otherwise it is labeled as unsuccessful.

Evaluation metrics such as average retrieval precision and average retrieval rate are utilized to determine retrieval efficiency of the proposed LMeQEP approach.

### 6.3.1   MIT VisTex Database

MIT VisTex repository possessing 40 images with varying texture content is used in experimental work. As discussed earlier, every image from database becomes query image. Evaluation metrics such as average retrieval precision and average retrieval rate are used to determine the effectiveness of the method.

Figure 6.3 compares various methods against the proposed method. Average recall rate is the evaluation measure considered for MIT VisTex database. It is evident that LMeQEP exhibits a substantial development in the performance when compared to related approaches. Figure 6.4 provides the query results of the proposed LMeQEP on MIT VisTex repository.

### 6.3.2   Corel-1k

Corel-1k database is used for experimentation. Figure 6.5 compares average retrieval precision of proposed method and other methods. LMeQEP method shows a substantial increase in average retrieval precision values as compared to other recent methods. Figure 6.6 compares precision with recall of different techniques on Corel-1k database. From Figures 6.5 and 6.6, it is evident that devised method surpasses other methods on Corel-1k repository. Figure 6.7 exhibits retrieval results from Corel-1k database for a query image.

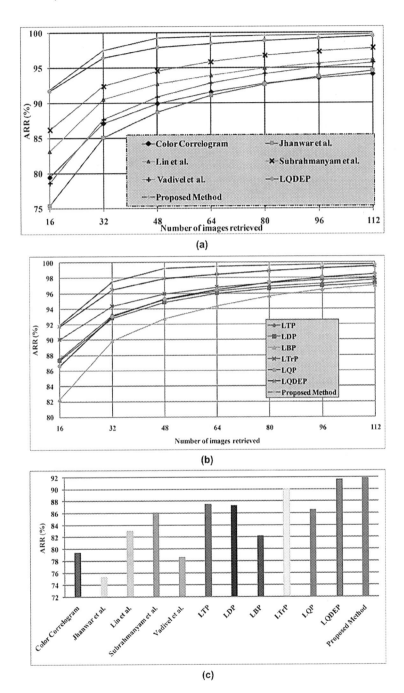

**FIGURE 6.3** Comparison of devised approach and other approaches in terms of average recall rate.

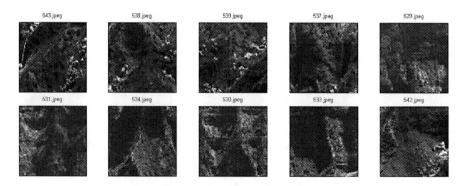

**FIGURE 6.4**    Results of retrieval made using the introduced method.

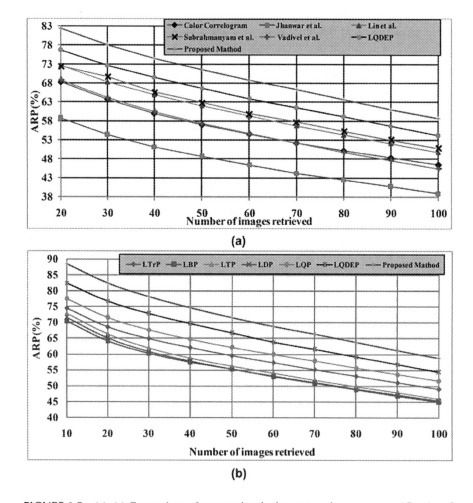

**FIGURE 6.5**    (a)–(c) Comparison of proposed and other approaches.          (*Continued*)

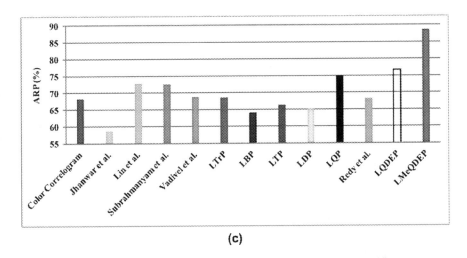

**(c)**

**FIGURE 6.5 (Continued)** (a)–(c) Comparison of proposed and other approaches.

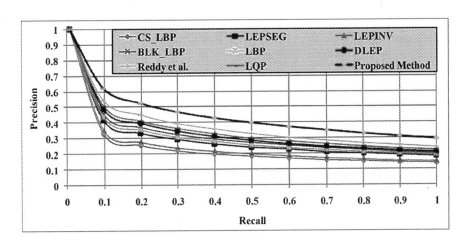

**FIGURE 6.6** Precision vs. recall comparison of proposed method and other methods.

**FIGURE 6.7** Query results of the experiment on Corel-1k repository.

## 6.4   CONCLUSION

An approach based on mesh concept, proposed for content-based image retrieval is presented in this chapter. LMeQEP extracts directional information from mesh structure formed from quantized geometric structure. Performance of the proposed method is determined by executing experiments on standard image repositories. Retrieval output depicts a momentous increase in evaluation metric as compared to many other approaches. In the next chapter, few more local extrema patterns are presented.

## SOLVED PROBLEMS

1. The given 12 bit image is converted to LMeQEP map. What is the range of intensity values of the image?
   **Ans:** The range of intensity values for a given 12 bit image is 0 to 4095.

2. Draw LMeQEP $HVDA_{11}$ geometric structure for the image of $11 \times 11$ size.

| 11 | 40 | 77 | 93 | 63 | 5 | 61 | 83 | 121 | 43 | 31 |
|----|----|----|----|-----|----|-----|-----|-----|-----|-----|
| 88 | 12 | 94 | 38 | 118 | 4 | 90 | 39 | 89 | 30 | 117 |
| 47 | 129 | 13 | 78 | 62 | 3 | 52 | 111 | 29 | 72 | 44 |
| 87 | 55 | 64 | 14 | 48 | 2 | 67 | 28 | 84 | 65 | 109 |
| 32 | 75 | 34 | 95 | 15 | 1 | 27 | 110 | 35 | 106 | 33 |
| 112 | 76 | 58 | 79 | 107 | P | 80 | 51 | 60 | 66 | 100 |
| 54 | 59 | 53 | 120 | 26 | 6 | 16 | 99 | 96 | 50 | 97 |
| 73 | 135 | 72 | 25 | 49 | 7 | 68 | 17 | 142 | 70 | 101 |
| 46 | 74 | 24 | 122 | 81 | 8 | 69 | 71 | 18 | 143 | 45 |
| 86 | 23 | 82 | 37 | 92 | 9 | 85 | 36 | 103 | 19 | 98 |
| 21 | 41 | 115 | 57 | 131 | 10 | 141 | 91 | 56 | 42 | 20 |

**Ans:** The $HVDA_{11}$ geometric structure is

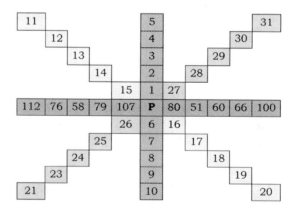

3.  Draw LMeQEP HVDA$_9$ geometric structure for image of 9×9 size.

| 12 | 94 | 38 | 118 | 4 | 90 | 39 | 89 | 30 |
|---|---|---|---|---|---|---|---|---|
| 129 | 13 | 78 | 62 | 3 | 52 | 111 | 29 | 72 |
| 55 | 64 | 14 | 48 | 2 | 67 | 28 | 84 | 65 |
| 75 | 34 | 95 | 15 | 1 | 27 | 110 | 35 | 106 |
| 76 | 58 | 79 | 107 | P | 80 | 51 | 60 | 66 |
| 59 | 53 | 120 | 26 | 6 | 16 | 99 | 96 | 50 |
| 135 | 72 | 25 | 49 | 7 | 68 | 17 | 142 | 70 |
| 74 | 24 | 122 | 81 | 8 | 69 | 71 | 18 | 143 |
| 23 | 82 | 37 | 92 | 9 | 85 | 36 | 103 | 19 |

**Ans:** The HVDA$_9$ geometric structure is

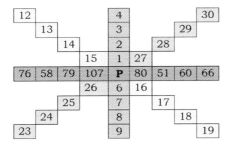

4.  Draw LMeQEP HVDA$_7$ geometric structure for the image of 7×7 size.

| 13 | 78 | 62 | 3 | 52 | 111 | 29 |
|---|---|---|---|---|---|---|
| 64 | 14 | 48 | 2 | 67 | 28 | 84 |
| 34 | 95 | 15 | 1 | 27 | 110 | 35 |
| 58 | 79 | 107 | P | 80 | 51 | 60 |
| 53 | 120 | 26 | 6 | 16 | 99 | 96 |
| 72 | 25 | 49 | 7 | 68 | 17 | 142 |
| 24 | 122 | 81 | 8 | 69 | 71 | 18 |

**Ans:** The HVDA$_7$ geometric structure is

5. Draw LMeQEP HVDA$_5$ geometric structure for the image of 5×5 size.

| 14 | 48 | 2 | 67 | 28 |
|----|-----|---|----|-----|
| 95 | 15 | 1 | 27 | 110 |
| 79 | 107 | P | 80 | 51 |
| 120 | 26 | 6 | 16 | 99 |
| 25 | 49 | 7 | 68 | 17 |

**Ans:** The HVDA$_5$ geometric structure is

## EXERCISES

1. Draw LMeQEP HVDA$_7$ geometric structure for the image of 7×7 size.

| 23 | 81 | 62 | 32 | 52 | 22 | 56 |
|----|----|-----|----|----|-----|----|
| 44 | 74 | 48 | 21 | 67 | 28 | 84 |
| 34 | 51 | 15 | 17 | 27 | 110 | 35 |
| 63 | 29 | 107 | P | 25 | 51 | 60 |
| 87 | 12 | 26 | 6 | 32 | 99 | 96 |
| 55 | 53 | 29 | 36 | 1 | 74 | 41 |
| 16 | 27 | 81 | 8 | 99 | 63 | 18 |

2. Draw LMeQEP HVDA$_5$ geometric structure for the image of 5×5 size.

| 4 | 8 | 2 | 7 | 8 |
|----|---|---|---|---|
| 9 | 5 | 1 | 2 | 0 |
| 7 | 7 | P | 8 | 5 |
| 12 | 6 | 6 | 6 | 9 |
| 2 | 9 | 7 | 5 | 1 |

3. Draw LMeQEP HVDA$_9$ geometric structure for the image of 9×9 size.

| 21 | 4 | 21 | 18 | 54 | 0 | 93 | 94 | 30 |
|----|----|----|----|----|----|----|----|----|
| 29 | 31 | 52 | 26 | 23 | 2 | 11 | 58 | 72 |
| 54 | 6 | 41 | 48 | 42 | 7 | 84 | 47 | 65 |
| 56 | 42 | 84 | 64 | 61 | 2 | 10 | 21 | 06 |
| 67 | 97 | 65 | 17 | P | 80 | 51 | 46 | 26 |
| 85 | 25 | 12 | 68 | 26 | 61 | 12 | 67 | 57 |
| 13 | 8 | 5 | 47 | 78 | 86 | 32 | 41 | 71 |
| 47 | 4 | 22 | 18 | 84 | 65 | 42 | 88 | 13 |
| 32 | 20 | 7 | 29 | 19 | 47 | 67 | 31 | 19 |

4. Draw LMeQEP HVDA$_3$ geometric structure for the image of 3×3 size. What is your observation?

| 5 | 1 | 2 |
|---|---|---|
| 7 | **P** | 8 |
| 9 | 7 | 5 |

5. Extract the 45° and 90° extrema patterns for the image of 5×5 size.

| 52 | 41 | 35 | 21 | 14 |
|----|----|----|----|----|
| 46 | 38 | 29 | 53 | 43 |
| 36 | 43 | C  | 41 | 92 |
| 26 | 33 | 27 | 38 | 73 |
| 50 | 48 | 29 | 30 | 52 |

# 7 Local Patterns for Feature Extraction

## 7.1 QUANTIZED NEIGHBORHOOD LOCAL INTENSITY EXTREMA PATTERNS FOR IMAGE RETRIEVAL

### 7.1.1 INTRODUCTION

P. Banerjee et al. (2018) introduced the local intensity patterns. It is an extended concept of Local Binary Patterns (LBP). Relationship among pixels in a neighborhood is used in this method. Major limitations in terms of the neighborhood and increased feature vector length influenced us to propose Quantized neighborhood local intensity extrema patterns for texture extraction. Structures in four directions, i.e. horizontal, vertical, diagonal and anti-diagonal are taken. In other words, the image is quantized in terms of number of gray levels. We use a 5 × 5 window in a quantized form. As shown in Figure 7.1, center pixel Pc is taken as reference. For a specific pixel $P_z$, number of adjacent neighbors can be either 4 or 2 depending on "z" value. Here, $Sn_z$ denotes the set of adjacent neighbors as shown in Equations (7.1) and (7.2). In the first method, mutual textural relationship between center pixel and neighboring elements is extracted. In our second method, relationship between different pixels of a neighborhood is used. More details of the proposed method are depicted in Figure 7.1.

$$Sn_z = \{P_{1+\mathrm{mod}(z+5,7)}, P_{1+\mathrm{mod}(z+6,9)}, P_{z+1}, P_{2z+9}, P_{1+\mathrm{mod}(z+2,8)}\} \forall z = 1,3,5,7c \qquad (7.1)$$

$$Sn_z = \{P_{z-1}, P_{2z+5}, P_{9+2\mathrm{mod}(z,8)}, P_{2z+7}, P_{\mathrm{mod}(2z+1,8)}\} \forall z = 2,4,6,8 \qquad (7.2)$$

In the first step, sign of a relative variation between any one neighbor of central pixel Pc and neighbors of the same specific pixel $Sn_z$ is identified. An N-bit pattern is obtained in which N denotes no. of elements in Snz. We use Equations (7.3) and (7.4) to obtain two different binary patterns one related to the central pixel and the other related to Pz.

$$Q_{1,z}(y) = sign\left(Sn_z(y), P_c\right) \text{where, } y = 1\, to\, N \qquad (7.3)$$

$$Q_{2,z}(y) = sign\left(Sn_z(y), P_c\right) \text{where, } y = 1\, to\, N \qquad (7.4)$$

$$sign(u,v) = \begin{cases} 1 & u \geq v \\ 0 & otherwise \end{cases} \qquad (7.5)$$

DOI: 10.1201/9781003123514-7

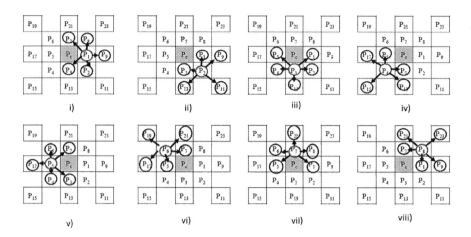

**FIGURE 7.1**    Illustration of adjacent neighbors for a quantized structure of size 5 × 5.

In the equation above, u and v are numbers. Logical XoR operation is performed to identify the structural change present in the specified neighborhood. Upon completing the XOR operation, an N bit pattern $E_z$ is extracted as per Equation (7.6)

$$E_z = xor\left(Q_{1,z}, Q_{2,z}\right) \qquad (7.6)$$

The count of 1s in the pattern helps in deciding the structural change. For any N bit patterns, total no. of positions of difference ranges from 0 to N. Here is ¼ (N) is fixed as threshold value, N = 4 and 2 when z ∈ odd and z ∈ even, respectively. Value of threshold is used in the calculation of a bit for $P_z$ according to Equation (7.7). For all eight neighbors, bit pattern is computed. Bit pattern map for center pixel is extracted using Equation (7.8). Subsequently, a histogram is built to represent the sign pattern.

$$Q\left(P_z, P_c\right) = \begin{cases} 1, & \#\left(E_z = 1\right) \geq \dfrac{1}{2}\left(N\right) \\ 0, & otherwise \end{cases} \qquad (7.7)$$

$$QNLIEP_{Sn}\left(P_c\right) = \sum_{z=1}^{8} 2^{z-1} \times Q\left(P_z, P_c\right) \qquad (7.8)$$

In the research work of Verma and Raman (2016) and Reddy and Reddy (2014), they have proved that magnitude plays an important role in local patterns. In the proposed method, we have taken magnitude into account and extracted another pattern. Statistical Dispersion is followed to explore a magnitude pattern from 3 × 3 kernel. To determine dispersion, absolute mean deviation is used. A threshold is fixed from mean deviation value of neighbors $P_z$. Equations (7.9) and (7.10) to extract the $QNLIEP_{Mg}$ are given below.

$$A_z = \frac{1}{A} \sum_{c=1}^{A} \left| Sn_z(c) - P_z \right| \tag{7.9}$$

$$Th_c = \frac{1}{8} \sum_{z=1}^{n} \left| P_z - P_c \right| \tag{7.10}$$

Here, $A_z$ is mean deviation around $z^{th}$ neighboring element of Pc from Snz, where $z = 1, 2, 3...., 8$. Calculation is repeated for all its neighbors. After determining the threshold, final bit of $z^{th}$ neighbor of $P_c$ is extracted. Comparison of $A_z$ with threshold value is done according to Equation (7.11). Final magnitude pattern corresponding to the center pixel is computed using Equation (7.12).

$$A((P_z, Th_c) = sign(A_z, Th_c) \tag{7.11}$$

$$QNLIEP_{Mg}(P_c) = \sum_{z=1}^{8} 2^{z-1} \times A(P_z, Th_c) \tag{7.12}$$

Histogram for the magnitude pattern is calculated in the same way as that of sign pattern. After obtaining the sign pattern and magnitude pattern, we concatenate to create the Quantized Local Neighborhood Intensity Extrema Pattern (QNLIEP) as shown in Equation (7.13).

$$HIST = \left[ HIST^{QNLEPL_{Sn}}, HIST^{QNLEPL_{Mg}} \right] \tag{7.13}$$

Illustration of the QNLIP pattern is provided in Figure 7.2.

A quantized structure example is shown in Figure 7.2(a). Central pixel is marked in blue color in Figure 7.2(b). For the example pattern in Figure 7.2(b), extraction of the pattern is illustrated in figures ((c)–(j)). Center pixel is shown in blue, pixels under consideration are colored in orange and neighboring elements are colored in pink. Initially, orange pixel and a group of pink pixels are compared. Binary value "0" or "1" is labeled according to comparison. As depicted in Figure 7.2(c), P1 and Pc are compared with P2, P3, P7, P8 and P9. Since Pc is less than P2, P3, P7, P8 and P9, we get a pattern 11111 and since Pc is lesser than P2 and P8 and Pc is greater than P3 and P7, the binary string 01001 is formed. Bit-wise XoR operation is performed between the binary strings 11111 and 01001, which results in a string 10110. In this case, three bits are different, which is more than threshold 2. Therefore, the bit meant for P1 is 0. Similarly, patterns are extracted from neighboring pixels. Binary values for remaining pixels are extracted in the same way. Accordingly, center pixel's sign pattern generated is 01010111. Mean deviation of neighborhood pixels about a specific pixel ($P_z$) is calculated. This value is used to compare with the threshold value.

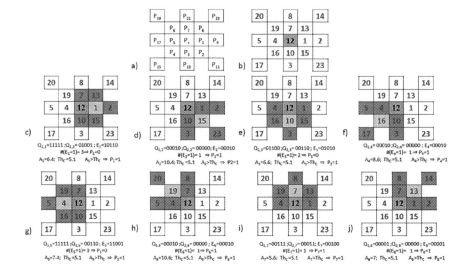

**FIGURE 7.2**   Illustration of quantized neighborhood local intensity extrema pattern. (a)–(b) Quantized structure highlighting the center and neighboring pixels; (iii)–(x) Calculation of sign and magnitude pattern calculation.

In the window taken, 12 is central pixel and P1 is 1. Value of mean deviation for an adjacent neighboring pixel P1 is calculated. A binary 1 or zero is assigned according to the difference between the threshold value and the mean deviation. If the magnitude pattern is greater than the threshold, binary 1 is assigned else 0. Calculation of magnitude pattern is depicted in (c)–(j). Accordingly, magnitude pattern generated is 1111111.

### 7.1.2   MAJOR ADVANTAGES OVER OTHER METHODS

Local patterns extract the structural arrangement of pixels from an image. LBP explores the textural information by comparing the central pixel and its neighbors at a time. Existing LNIP patterns are limited to 3×3 windows. The process becomes more complex when it is extended to next higher sizes. In contrast, the proposed feature vector depends on quantizing the image. The proposed method can be extended to any size of the window such as 7×7 or 9×9. Main advantages of quantizing are as follows. Size of the window can be extended to 5×5 or 9×9 and so on to any level. By quantizing an image of size, for instance, in a 5×5 window, 17 pixels and the relationship among them is extracted instead of 25 pixels. Feature vector length is reduced and thus complexity becomes less as compared to the full-size window.

### 7.1.3   FRAMEWORK OF THE PROPOSED RETRIEVAL SYSTEM

Detailed block diagram of retrieval system is shown in Figure 7.3, followed by algorithm. An image is considered as input and a feature vector is obtained in the output.

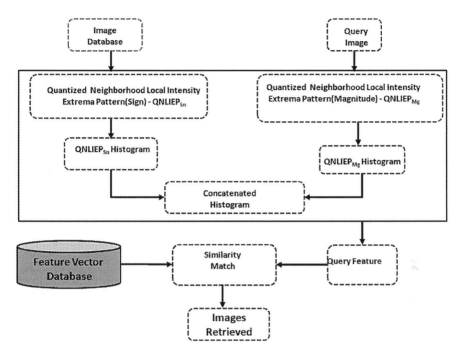

**FIGURE 7.3** Schematic diagram of the proposed approach.

## ALGORITHM

1) Get image from database and make it gray, if it is in color form.
2) Extract $QNLIEP_{Sn}$ to build histogram.
3) Extract $QNLIEP_{Mg}$ to build histogram.
4) Create a final feature vector by concatenating histograms in step 2 and step 3.
5) Take a query image as input.
6) Derive the sign patterns and magnitude patterns as given in steps 2 and 3.
7) Calculate the similarity for query and database image.
8) Arrange the similarity indices in descending order from the highest to the lowest.
9) Apply the evaluation metrics such as precision and recall.

### 7.1.4 IMAGE SIMILARITY MEASUREMENT

For any retrieval system, apart from feature vector extraction, similarity measure equally has a significant role. Subsequent to calculation of feature vector, distance between a query and database images is computed to determine the level of similarity. Five different types of distance metrics are used for similarity calculation.

a) Euclidian distance: $\quad distMesr_{S,t_z} = \left( \sum_{k=1}^{o} \left| D_s^{\ z}(k) - D_{t_z}(k) \right|^2 \right)^{1/2}$    (7.14)

b) dl distance $\quad distMesr_{S,t_z} = \sum_{k=1}^{o} \left| \dfrac{D_s^{\ z}(k) - D_{t_z}(k)}{1 + D_s^{\ z}(j) - D_{t_z}(k)} \right|$    (7.15)

c) Canberra distance $\quad distMesr_{s,t_z} = \sum_{k=1}^{o} \left| \dfrac{D_s^{\ z}(k) - D_{t_z}(k)}{1 + D_s^{\ z}(k) - Dt_z(k)} \right|$    (7.16)

d) Chi-square distance $\quad distMesr_{s,t_z} = \dfrac{1}{2} \sum_{k=1}^{o} \left| \dfrac{D_s^{\ z}(k) - D_{t_z}(k)^2}{D_s^{\ z}(k) - D_{t_z}(k)} \right|$    (7.17)

e) Manhattan distance $\quad distMesr_{s,t_z} = \sum_{k=1}^{o} \left| D_s^{\ z}(k) - D_{t_z}(k) \right|$    (7.18)

## 7.1.5 EXPERIMENTAL RESULTS AND DISCUSSION

Performance of retrieval framework primarily relies on effective retrieval of similar images. Algorithm has to be tested with varying content of imagery data and the most similar images are to be arranged in descending order of similarity. For this purpose, we have used precision and recall as evaluation metrics. Definition of evaluation metrics and the related equations are specified below in the equations below.

$$Pr_{avg}(C) = \frac{1}{I} \sum_{x=1}^{I} Pr_x \qquad (7.19)$$

$$Re_{avg}(C) = \frac{1}{I} \sum_{r=1}^{I} Re_r \qquad (7.20)$$

Total precision and recall are calculated using the Equations (7.21) and (7.22) given below.

$$Ps_{total}(C) = \frac{1}{k} \sum_{c=1}^{K} ps_{avg}(C) \qquad (7.21)$$

$$Re_{total}(C) = \frac{1}{k} \sum_{c=1}^{K} Re_{avg}(C) \qquad (7.22)$$

### 7.1.5.1  Database: 1

Corel database contains 10000 images. Entire repository is classified into 100 categories with varying content. Many researchers consider the Corel database as an appropriate tool to test the effectiveness of their algorithms mainly due to the content of images. In a specific category, each image is treated as query image. Precision and recall are used as evaluation metrics t test for the effectiveness of the proposed algorithm. QNLIEP supersedes LTrDP by 4.9 %, LEPSEG by 10.9 %, CSLBP 10.8 by%, LBP by19.2 %, LEPINV by 6.4 %, LNDP by 4.4 %, and LNIP by 2.3 % in terms of average retrieval rate. Comparison of recent methods w.r.t. the proposed method is given in Figure 7.4. Images retrieved for a given query are shown in Figure 7.5.

### 7.1.5.2  Database: 2

ImageNet-25k is used to test efficiency of the proposed method. Similar to the procedure in experiment 1, each image of database becomes a query. Precision and recall are evaluation metrics used here to determine the effectiveness of algorithm. LTrDP by 2.05%, LEPSEG by 5.5%, CSLBP by4.7%, LBP by 11.18%, LEPINV by 4.87%, LNDP by 3.79%, LNIP by 2.05%. Comparison of recent methods w.r.t. the proposed method is given in Figure 7.6. Images retrieved for a given query are shown in Figure 7.7.

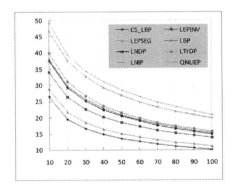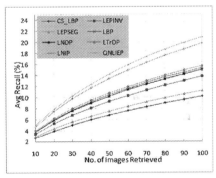

**FIGURE 7.4**    Evaluation metrics calculated on database 1.

**FIGURE 7.5**    Retrieval results for a query from the database 1.

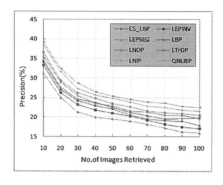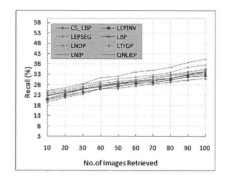

**FIGURE 7.6**    Evaluation metrics calculated on database 2.

**FIGURE 7.7**    Retrieval results for a query from database 2.

### 7.1.5.3    Database: 3

Brodatz dataset consists of 112 images of size 640×640. Each Image is divided into 25 images, each of size 128×128. Therefore, there exist 112 classes. Each class has 25 images. Efficiency in terms of precision and recall are used to test the effectiveness of method. In the experiment, initially 25 images are retrieved. In the next step, it is increased in parts of 5. The proposed method surpasses the other methods such as LTrDP by 3.73%, LEPSEG by 10.86%, CSLBP by 21.54 %, LBP by 27%, LEPINV by 20.87%, LNDP by 4.37 % and LNIP by 2.08 % against the average retrieval rate as depicted in the table below. Comparison of recent methods w.r.t. the proposed method is given in Figure 7.8. Images retrieved for a given query are shown in Figure 7.9.

### 7.1.5.4    Database: 4

Salzburg Database has 7616 images with the size of 128×128. There exist 476 categories. Each category has 16 images. Every image of the database is considered as the query image. Initially, only 16 are considered. Number of images retrieved for each category for this experiment is initially considered as 16 and the count is increased in steps of 16. QNLIEP method shows significant improvement in average retrieval rate (ARR) as compared to many popular methods. QNLIEP surpasses the

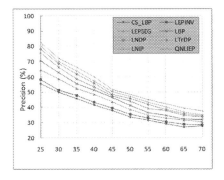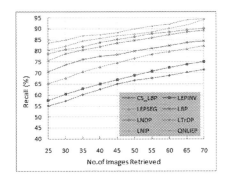

**FIGURE 7.8**    Evaluation metrics calculated on database 3.

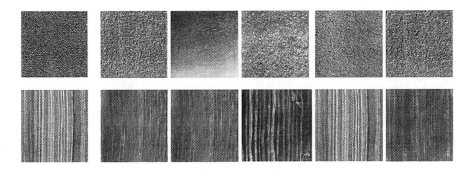

**FIGURE 7.9**    Retrieval results for a query from the database 3.

other methods like LTrDP by 2.45 %, LEPSEG by 11.55%, CSLBP by 18.45%, LBP by 26.45%, LEPINV by 21.3 %, LNDP by 3.84 % and LNIP by 3.62 %. Comparison of recent methods w.r.t. the proposed method is given in Table 7.1.

Figure 7.10 Images retrieved for a given query are shown in Figure 7.11.

Comparison in terms of complexity, retrieval time is provided in Table 7.1, Table 7.2 and Table 7.3.

## TABLE 7.1
### Feature Vector Length and CPU Time Values of Relevant Methods

|  | CPU Time | Time for Feature Extraction | Feature Vector Length |
|---|---|---|---|
| **LTrDP** | 0.0332 | 0.0305 | 768 |
| **LEPSEG** | 0.0330 | 0.0371 | 512 |
| **CS_LBP** | 0.0322 | 0.0196 | 16 |
| **LBP** | 0.0325 | 0.0192 | 256 |
| **LEPINV** | 0.0323 | 0.0720 | 72 |
| **LNDP** | 0.0321 | 0.0369 | 64 |
| **LNIP** | 0.0345 | 0.0348 | 512 |
| **QNLIEP (Proposed)** | 0.0281 | 0.0310 | 256 |

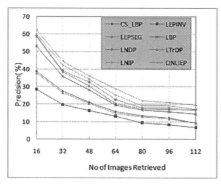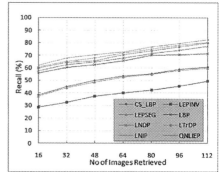

**FIGURE 7.10**   Evaluation metrics calculated on database 4.

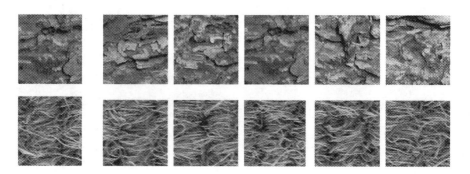

**FIGURE 7.11**   Retrieval results for a query from the database 4.

**TABLE 7.2**

**Average Retrieval Rates**

|  | Corel | ImageNet | Brodatz | Salzburg |
|---|---|---|---|---|
| **LTrDP** | 50.8 | 43.79 | 75.97 | 58.36 |
| **LEPSEG** | 47.6 | 40.33 | 69.04 | 49.26 |
| **CS_LBP** | 46.9 | 41.14 | 58.36 | 42.36 |
| **LBP** | 36.5 | 34.66 | 52.9 | 34.36 |
| **LEPINV** | 49.3 | 41.37 | 59.03 | 39.51 |
| **LNDP** | 51.3 | 42.05 | 75.53 | 56.97 |
| **LNIP** | 53.4 | 43.79 | 77.08 | 57.19 |
| **QNLIEP (Proposed)** | **55.7** | **45.84** | **79.9** | **60.81** |

From Table 7.1, it is evident that the proposed method shows improvement over other methods in terms of CPU time, feature vector length and time for retrieval.

From the Table 7.2, it is evident that the proposed method shows improvement over other methods in terms of average retrieval rate.

**TABLE 7.3**
**Comparison of Different Distance Metrics**

|            | Corel | ImageNet | Brodatz | Salzburg |
|------------|-------|----------|---------|----------|
| Euclidian  | 12.3  | 14.58    | 75.34   | 54.21    |
| d1         | 14.3  | 15.63    | 83.83   | 62.81    |
| Canberra   | 11.8  | 12.04    | 75.27   | 51.24    |
| Chi-square | 13.8  | 13.65    | 81.25   | 59.27    |
| Manhattan  | 13.5  | 13.11    | 77.200  | 57.67    |

### 7.1.6 CONCLUSION

A novel feature descriptor, quantized neighborhood local intensity extrema pattern is presented for retrieval. In contrast to many of the existing methods, local relationship among a group of pixels is extracted at a time instead of comparing limited number of pixels. Sign as well as magnitude patterns are derived. Local extrema is also calculated to extract more discriminating information when more than two–three pixels are present. Precision and recall result show considerable increase in performance.

In future, the number of pixels under consideration can be increased further to extend this work. Further, the method can be tested with many face recognition databases.

## 7.2 MAGNITUDE DIRECTIONAL LOCAL EXTREMA PATTERNS

### 7.2.1 INTRODUCTION

Content-based image retrieval (CBIR) became an eye catching area for many researchers across the world to resolve the issues of existing methods. In this method, the visual contents like texture, color, shape, etc., are taken out to make the signature or the vector. The resemblance of query image with database image is considered to explore similar images out of the dataset. In fact, capability of CBIR system mainly relies upon the method to extract characteristics like texture color, shape, layout, etc. A. W. M. Smeulders et al. (2000), Veltkamp and Tanase (2002), M. L. Kherfi et al. (2004).

Classification and segmentation became very critical in analyzing the texture of an image. In the work Tuceryan and Jain et al. (1998), use of texture property to classify image was mentioned. Arivazhagan and Ganesan (2003) presented an approach to classify the texture that uses wavelet transform. Wavelet Packet Frame Decomposition (WPFD) and Gaussian Mixture Model (GMM) were specified in S. C. Kim and Kang (2007) to classify the texture and segmentation.

Gabor wavelets played an important role in classifying the texture for rotation invariant feature because of their closeness to the human visual system S. Arivazhagan and Ganesan (2006).

### 7.2.1.1 Contribution

Existing Magnitude Directional Local Extrema Patterns (MDLEP) extracts directional and magnitude data of edges as per the minima or maxima in vertical, horizontal, diagonal, anti-diagonal directions of image. In this chapter, a new method is proposed that combines color feature and MDLEP to improve the output of existing MDLEP. The chapter is arranged as follows: Different kinds of local patterns that are related to the work are reviewed in Section 7.2.2. Section 7.2.3 mentions the proposed approach for image retrieval. Section 7.2.4 shows results and the discussions and conclusions are provided in Section 7.2.5.

### 7.2.1.2 Review of Related Work

Zhang et al. (2010a) developed local derivative pattern by taking $n^{th}$ order LBP. Subrahmanyam et al. (2012) created an operator called Directional Local Extrema Pattern (DLEP) as a descriptor in texture analysis and classification. V. B. Reddy and Reddy (2014) enhanced the DLEP by taking magnitude into consideration. The MDLEP is different from the available LBP and modifications in taking out directional data.

### 7.2.2 Different Types of Local Patterns

### 7.2.2.1 Local Binary Pattern

T. Ojala et al. (2002) introduced LBP operator. In LBP, gray-scale value of center pixel is assumed as maximum level, difference in the value of center pixel and surrounding neighbors is considered to label a 0 or 1. Same procedure is followed till all elements around the center pixel get covered in the process.

$$LBP_{X,Y} = \sum_{p=0}^{p-1} y\left(x_p - x_c\right)2^p, y\left(c\right) = \begin{cases} 1 & c \geq 0 \\ 0 & c < 0 \end{cases} \tag{7.23}$$

where $x_c$ is gray value of center pixel and $x_p$ represents the intensity value of X equally spanned pixels on a circle of radius Y. For example, for the image given below, the pattern is 01110110.

| 22 | 34 | 49 |
|----|----|----|
| 14 | 27 | 65 |
| 71 | 58 | 16 |

### 7.2.2.2 Local Directional Pattern

It is based on the LBP which uses edge information of neighboring pixels to code texture of image. It labels an 8-bit code to every pixel in image.
    Value of 1 or 0 is coded based on the existence of an edge.

$$LDP_n = \sum_{i=1}^{8} g_i\left(m_i - m_k\right) * 2^i, g_i\left(x\right) = \begin{cases} 1, x \geq 0 \\ 0, x < 0 \end{cases} \tag{7.24}$$

### 7.2.2.3 Directional Local Extrema Patterns

Principle of LBP was utilized by Subrahmanyam et al. (2012) to design a novel descriptor named DLEP. In this method, two neighboring pixel intensities of one direction are compared to value of center pixel to code 0 or 1. It describes the structure of local texture based on center pixel's extrema. Maxima and minima values of four directions can be obtained by calculating the difference between the center element and all neighbors.

The calculation is mentioned in Equation (7.25).

$$M'(x_i) = M(x_c) - M(x_i); i = 1, 2, ....8 \qquad (7.25)$$

The local extrema's calculation is done according to the equations below.

$$\hat{M}_\beta(xc) = Y_3\left(M'(x_i) * M'(x_{j+4})\right); j = (1 + \beta / 45) \qquad (7.26)$$
$$\forall_\beta = 0°, 45°, 90°, 135°$$

$$f_3\left(I'(g_j), I'(g_{j+4})\right) = \begin{cases} 1 & I'(g_j) X I'(g_{j+4}) \geq 0 \\ 0 & \text{else} \end{cases}$$

$$Y_3\left(M'(x_j), M'(x_{j+4})\right) = \begin{cases} 1 & M'(x_j) * M'(x_{j+4}) \geq 0 \\ 0 & \text{else} \end{cases} \qquad (7.27)$$

The DLEP is computed as ($\beta = 0°, 45°, 90°$ and $135°$) follows:

$$\text{DLEP}\left(M(x_c)\right) | \beta = \{M_\beta(x_c); \hat{M}_\beta(x_i); \hat{M}_\beta(x_z); .... \hat{M}_\beta(x_8) \qquad (7.28)$$

The details about DLEP are given in Figure 7.12. Consequently, the image is changed DLEP output having the values 0 to 511.

In the next level, DLEP, image is denoted by getting a histogram as per the Equation (7.29).

$$H_{DLEP|\beta}(l) = \sum_{m=1}^{Z_1} \sum_{n=1}^{Z_2} Y_2\left(DLEP(m,n)\big|_\alpha, \ell\right); \qquad (7.29)$$
$$\ell \in (0, 511)$$

Here the $Z_1$. $Z_2$ is the dimension. Collection of DLEP data for a pixel at the center in the procedure for calculation of DLEP for center pixel marked in blue color is given Figure 7.12. Information in directions is taken out using local difference between center pixel and neighbors.

For example, DLEP of $90°$ direction for pixel highlighted with blue color is given in Figure 7.13. For a pixel of value 36, two neighbor pixels are moving away in values. Hence, pattern is assigned 1. Similarly, remaining bits of DLEP are collected making the final outcome as 110011110. In this way, the DLEPs are computed in $0°, 45°$ and $135°$ directions (Figure 7.14).

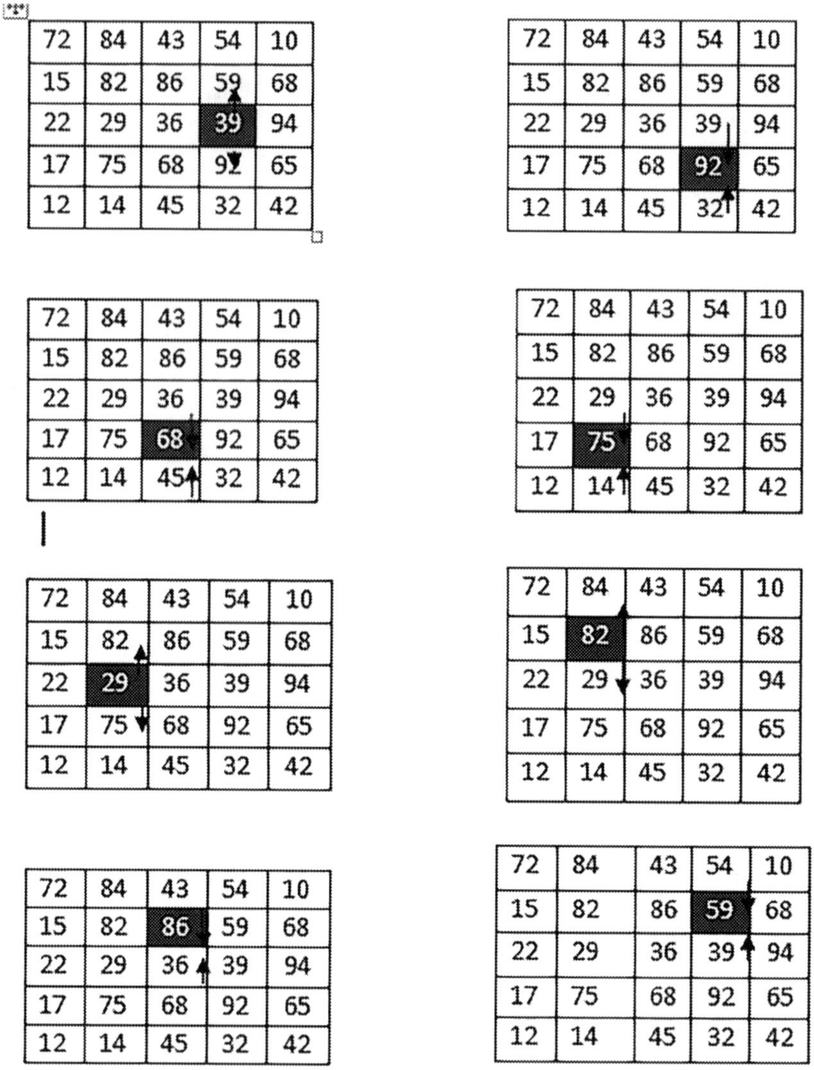

| 72 | 84 | 43 | 54 | 10 |
|----|----|----|----|----|
| 15 | 82 | 86 | 59 | 68 |
| 22 | 29 | 36 | **39** | 94 |
| 17 | 75 | 68 | 9**2** | 65 |
| 12 | 14 | 45 | 32 | 42 |

| 72 | 84 | 43 | 54 | 10 |
|----|----|----|----|----|
| 15 | 82 | 86 | 59 | 68 |
| 22 | 29 | 36 | 39 | 94 |
| 17 | 75 | 68 | **92** | 65 |
| 12 | 14 | 45 | 32 | 42 |

| 72 | 84 | 43 | 54 | 10 |
|----|----|----|----|----|
| 15 | 82 | 86 | 59 | 68 |
| 22 | 29 | 36 | 39 | 94 |
| 17 | 75 | **68** | 92 | 65 |
| 12 | 14 | 45 | 32 | 42 |

| 72 | 84 | 43 | 54 | 10 |
|----|----|----|----|----|
| 15 | 82 | 86 | 59 | 68 |
| 22 | 29 | 36 | 39 | 94 |
| 17 | **75** | 68 | 92 | 65 |
| 12 | 14 | 45 | 32 | 42 |

| 72 | 84 | 43 | 54 | 10 |
|----|----|----|----|----|
| 15 | 82 | 86 | 59 | 68 |
| 22 | **29** | 36 | 39 | 94 |
| 17 | 75 | 68 | 92 | 65 |
| 12 | 14 | 45 | 32 | 42 |

| 72 | 84 | 43 | 54 | 10 |
|----|----|----|----|----|
| 15 | **82** | 86 | 59 | 68 |
| 22 | 29 | 36 | 39 | 94 |
| 17 | 75 | 68 | 92 | 65 |
| 12 | 14 | 45 | 32 | 42 |

| 72 | 84 | 43 | 54 | 10 |
|----|----|----|----|----|
| 15 | 82 | **86** | 59 | 68 |
| 22 | 29 | 36 | 39 | 94 |
| 17 | 75 | 68 | 92 | 65 |
| 12 | 14 | 45 | 32 | 42 |

| 72 | 84 | 43 | 54 | 10 |
|----|----|----|----|----|
| 15 | 82 | 86 | **59** | 68 |
| 22 | 29 | 36 | 39 | 94 |
| 17 | 75 | 68 | 92 | 65 |
| 12 | 14 | 45 | 32 | 42 |

**FIGURE 7.12**  Description of DLEP for 5 × 5 size.

### 7.2.2.4 Magnitude Directional Local Extrema Patterns

V. B. Reddy and Reddy (2014) presented a technique to increase performance by collecting magnitudes of the Local patterns. MDLEP is collected according to the equation below.

$$\hat{I}_{M\beta}(xc) = Y_3\left(I'(x_i) * I'(x_{j+4})\right); j = (1 + \beta / 45)$$
$$\forall_\beta = 0°, 45°, 90°, 135°$$

(7.30)

| | $0_{\{27\}}$ | $1_{\{29\}}$ | $2_{\{80\}}$ | $3_{\{87\}}$ | $4_{\{88\}}$ | $5_{\{13\}}$ | $6_{\{78\}}$ | $7_{\{85\}}$ | $8_{\{63\}}$ | DLEP |
|---|---|---|---|---|---|---|---|---|---|---|
| P $(0^\circ)$ | 0 | 0 | 0 | 0 | 1 | 1 | 0 | 1 | 0 | 26 |
| Q$(45^\circ)$ | 1 | 0 | 0 | 1 | 1 | 1 | 1 | 0 | 1 | 317 |
| R$(90^\circ)$ | 1 | 1 | 0 | 0 | 1 | 1 | 1 | 1 | 0 | 415 |
| S$(135^\circ)$ | 1 | 1 | 0 | 0 | 0 | 1 | 1 | 1 | 0 | 398 |

**FIGURE 7.13** Calculation of DLEP in $90^\circ$.

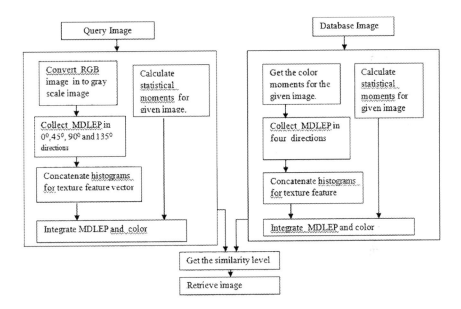

**FIGURE 7.14** The proposed approach for image retrieval.

$$Y_4\left(I'\left(x_j\right),I'\left(x_{j+4}\right)\right)=\begin{cases}1 & \mathrm{abs}\left(\,I'\left(x_j\right)\right)+\mathrm{abs}\left(I'\left(x_{j+4}\right)\right)\ge \mathrm{Thrs}\\ 0 & \mathrm{else}\end{cases} \qquad (7.31)$$

$$Thrs=\frac{1}{Z_1\times Z_2}\sum_{b=1}^{z_1}\sum_{c=1}^{z_2}\left(abs\left(I'\left(x_j\right)\Big|_{(b,c)}\right)+abs\left(I'\left(x_{j+4}\right)\Big|_{(b,c)}\right)\right) \qquad (7.32)$$

The MDLEP in $0^\circ$, $45^\circ$, $90^\circ$ and $135^\circ$ directions is defined as

$$MDLEP\left(I\left(x_c\right)\right)\big|\beta=\left\{\hat{I}_{MB}\left(x_c\right);\hat{I}_{MB}\left(x_1\right);\hat{I}_{MB}\left(x_2\right);\ldots\hat{I}_{MB}\left(x_8\right)\right\} \qquad (7.33)$$

Subsequent to the MDLEP calculation, entire image is shown by a histogram according to Equation (7.30).

### 7.2.3 The Proposed CMDLEP System

**ALGORITHM**

1. Calculate color moments for given image and convert RGB into gray-scale image.
2. Compute the local extrema in 0°, 45°, 90° and 135° directions.
3. Compute the MDLEP information in all four directions as per step 2.
4. Get histogram of MDLEP obtained from step 3 and join to create feature vector.
5. Combine the two features to form a feature vector used in the process of retrieval.

*Query Matching*

Once the features are extracted, the feature vector of query image is formed. Similarly, feature vectors of all images from database are collected. To recognize relevant image to query image, the distance between query image and repository images is calculated.

**TABLE 7.4**

**Results for Various Categories of the Database (in Table 7.1)**

| Category | Existing MDLEP | MDLEP+ color Feature | Category | MDLEP | MDLEP+ color Feature |
|---|---|---|---|---|---|
| Africans | 61.3 | 64.4 | Africans | 39.25 | 43.7 |
| Beach | 51.25 | 53.7 | Beach | 33.82 | 37.5 |
| Building | 57.85 | 62.6 | Building | 31.96 | 36.6 |
| Buses | 94.4 | 98.4 | Buses | 73.57 | 77.9 |
| Dinosaur | 97.85 | 99.1 | Dinosaur | 90.28 | 94.5 |
| Elephant | 48.9 | 64.8 | Elephant | 30.53 | 34.7 |
| Flower | 89.1 | 93.5 | Flower | 69.32 | 77.8 |
| Horse | 66.2 | 79.4 | Horse | 36.16 | 45.4 |
| Mountain | 39.4 | 48.5 | Mountain | 29.35 | 34.1 |
| Food | 75.35 | 90 | Food | 45.3 | 43.5 |
| Average Precision (%) | 68.16 | 75.44 | Average Recall (%) | 47.954 | 52.57 |

### 7.2.4 Experimental Results

Capability of the proposed method is tested with Corel-1K database. Precision (Pr) and recall (Re) values are calculated according to the equations below (Figures 7.15 and 7.16).

$$Pr = \frac{\text{Number of relevant images retrieved}}{\text{No.of images retrieved}}$$

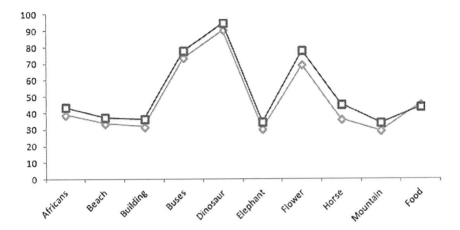

**FIGURE 7.15** Average precision of MDLEP (dark grey) and CMDLEP (light grey).

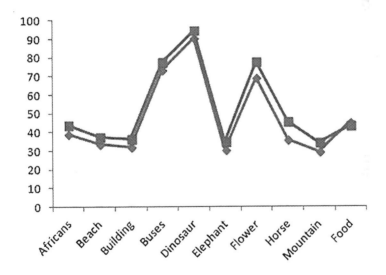

**FIGURE 7.16** Average recall of MDLEP (red) and CMDLEP (blue).

$$Re = \frac{\text{Number of relevant images retrieved}}{\text{Number of relevant images in the database}}$$

## 7.2.5 CONCLUSION

It is evident that proposed method is outperforming the recent MDLEP in terms of precision and recall. Combination of color and MDLEP is exploring more information existing in an image as compared to LBP and similar local pattern-based feature vectors.

## 7.3  COMBINATION OF CDLEP AND GABOR FEATURES

### 7.3.1  INTRODUCTION

Many techniques to classify and segment the texture can be found in the literature based on statistical analysis and structural methods in R. M. Haralick (1979). In R. M. Haralick et al. (1973a), the use of texture feature for classification of images was discussed. Arivazhagan and Ganesan (2003) proposed texture classification using wavelet transform. In S. C. Kim and Kang (2007) texture classification and segmentation was proposed using wavelet packet frames and Gaussian mixture model. Gabor wavelets were used in texture classification for rotation invariant features Arivazhagan and Ganesan (2006).

Gabor filters have been used in the field of Image processing and texture analysis A. A. Ursani et al. (2007) for many years. It is a linear, bandpass filter which is similar and close to human visual system. It gives the spatial frequency information.

#### 7.3.1.1  Contribution

The existing DLEP derives the directional edge information based on local extrema in 0°, 45°, 90° and 135° directions of an image. In this chapter, we propose a combination of features such as color DLEP and Gabor to improve the performance of the existing DLEP. The organization of this chapter is as follows: Section 7.3.2 explains Gabor feature. Section 7.3.3 covers the proposed work for retrieval system. Section 7.3.4 contains the results. The conclusions are given in Section 7.3.5.

#### 7.3.1.2  Related Work

A method based on LBP was introduced by Ojala et al. (2002) and the concept of LBP was extended to face recognition and other applications in T. Ahonen et al. (2006), G. Zhao and Pietikainen (2007). However, LBP has the limitation of rotational invariance in classifying the texture present in an image. Local derivative pattern by considering the nth order LBP was proposed by Zhang et al. (2010b). Subrahmanyam et al. (2012) proposed DLEP as a feature vector for texture analysis of an image. An improvement to DLEP was proposed by Koteswara Rao et al. (2015, 2016). The DLEP is different from the existing local patterns and other extensions in terms of directional information.

### 7.3.2  GABOR FEATURE

The Gabor filter is found to be efficient for text representation and discrimination. The representation of 2D Gabor filter is as specified below (Figures 7.17–7.20).

$$\psi_{f,\theta}(x,y) = \exp\left[-\frac{1}{2}\left\{\frac{x^2\theta_n}{\sigma^2 x} + \frac{y^2\theta_n}{\sigma^2 y}\right\}\right]\exp\left(2\pi fx\theta_n\right)$$

Here, S is the central frequency of sinusoidal plane and $\theta$ is the orientation of $xy$ plane.

$$\begin{bmatrix} x\theta_n \\ y\theta_n \end{bmatrix} = \begin{bmatrix} \sin\theta_n & \cos\theta_n \\ -\cos\theta_n & \sin\theta_n \end{bmatrix}\begin{bmatrix} x \\ y \end{bmatrix}(n-1)$$

| | $0_{(27)}$ | $1_{(29)}$ | $2_{(80)}$ | $3_{(87)}$ | $4_{(88)}$ | $5_{(13)}$ | $6_{(78)}$ | $7_{(85)}$ | $8_{(63)}$ | DLEP |
|---|---|---|---|---|---|---|---|---|---|---|
| P (0°) | 0 | 0 | 0 | 0 | 1 | 1 | 0 | 1 | 0 | 26 |
| Q(45°) | 1 | 0 | 0 | 1 | 1 | 1 | 1 | 0 | 1 | 317 |
| R(90°) | 1 | 1 | 0 | 0 | 1 | 1 | 1 | 1 | 0 | 415 |
| S(135°) | 1 | 1 | 0 | 0 | 0 | 1 | 1 | 1 | 0 | 398 |

(matrix:)
51 33 69 75 57
19 78 85 63 12
36 13 27 29 42
48 88 87 80 65
11 53 95 91 84

**FIGURE 7.17**  Illustration of DLEP for 3 × 3 pattern.

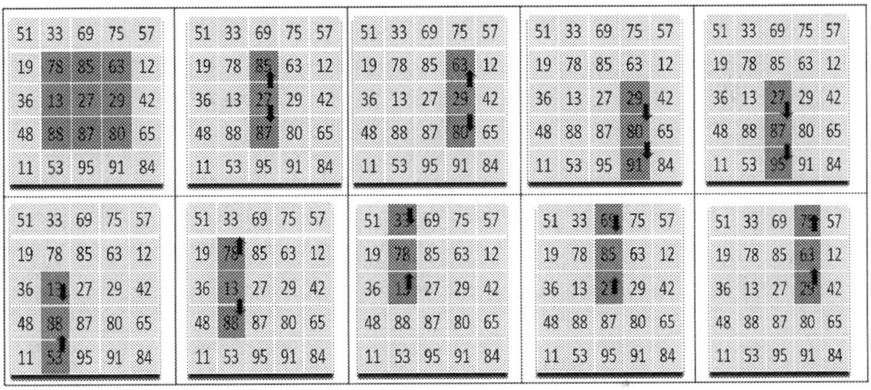

**FIGURE 7.18**  Example to compute DLEP in 90° direction (110011110).

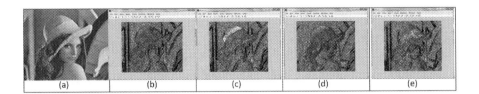

**FIGURE 7.19**  Example of DLEP maps: (a) Sample Image; (b) 0°; (c) 45°; (d)90°; (e) 135°.

$\theta_n$ is the rotation of xy plane by $\theta_n$ angle results Gabor filter at the orientation $\theta_n$.

$$angle\ \theta_n = \frac{\pi}{p}$$

where n=1,2,…p, p∈n and p is the orientation.

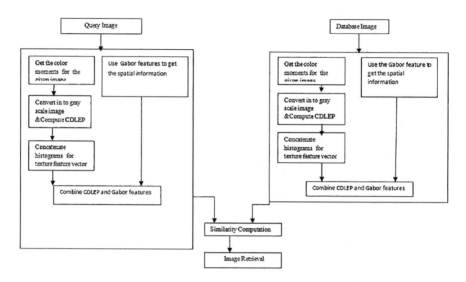

**FIGURE 7.20**    The proposed CDLEP+Gabor for content-based Image retrieval.

### 7.3.3   THE PROPOSED GABOR CDLEP SYSTEM

**THE PROPOSED ALGORITHM**

1. Compute the color moments for the given image and then convert RGB image into gray-scale image.
2. Use Gabor filter to get the spatial information.
3. Compute the local extrema in $0°$, $45°$, $90°$ and $135°$ directions.
4. Compute the CDLEP patterns in four directions mentioned in step 2.
5. Get a histogram for the DLEP patterns obtained in step 3 and concatenate to get the texture feature vector.
6. Combine these two features to get a feature vector that can be used in image retrieval.

*Query Matching*

Once the features are extracted, the feature vector for query image is obtained. In the same manner, feature vectors for all images in the database are also calculated. To select the more relevant image to the query image, the distance between query image and database images is calculated.

### 7.3.4   EXPERIMENTAL RESULTS

Performance of the CDLEP is evaluated on standard corel-1k database [16]. The precision (P) and recall (R) values are computed as per the relationship mentioned here under.

**TABLE 7.5**

**Comparison of Precision and Recall Values for DLEP and CDLEP**

| Category | Existing DLEP | CDLEP+ Gabor Feature | Category | DLEP | CDLEP+ Gabor Feature |
|---|---|---|---|---|---|
| Africans | 69.3 | 80 | Africans | 39.7 | 41 |
| Beach | 60.5 | 80 | Beach | 37.3 | 39 |
| Building | 72.0 | 100 | Building | 34.9 | 41 |
| Buses | 97.9 | 90 | Buses | 74.1 | 67 |
| Dinosaur | 98.5 | 100 | Dinosaur | 88.0 | 85 |
| Elephant | 55.9 | 80 | Elephant | 29.0 | 34 |
| Flower | 91.9 | 100 | Flower | 70.8 | 83 |
| Horse | 76.9 | 100 | Horse | 41.7 | 42 |
| Mountain | 42.7 | 60 | Mountain | 29.0 | 30 |
| Food | 82.0 | 90 | Food | 47.0 | 43 |
| Average Precision (%) | 74.8 | 88.0 | Average Recall (%) | 49.16 | 50.5 |

Comparison of precision values for DLEP and CDLEP

Comparison of recall values for DLEP and CDLEP

$$P = \frac{\text{No.of relevant images retrieved}}{\text{No.of images retrieved}}$$

$$R = \frac{\text{No.of relevant images retrieved}}{\text{No.of images in the database}}$$

The top ten results for different categories of the database are shown in Table 7.5. The comparisons in terms of average precision and recall are given in the graph given below (Figures 7.21–7.22).

### 7.3.5 CONCLUSION

It has been proved that the precision and recall values of the proposed method are better than the existing directional patterns. This work can be further extended by calculating the magnitudes of the pixels in each direction.

## 7.4 LEMP: A ROBUST IMAGE FEATURE DESCRIPTOR FOR RETRIEVAL APPLICATIONS

### 7.4.1 INTRODUCTION

In CBIR the extraction of image features is the primitive features. Effectiveness of this system primarily depends on feature extraction method. Key features of any image are color, shape, texture and layout, etc. Since the user captures a photograph is various conditions, there exists no single way of image representation. Survey on CBIR is

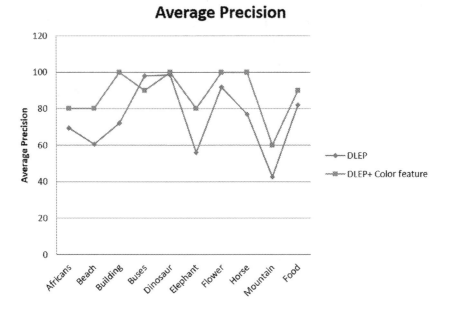

**FIGURE 7.21**   Category-wise performance in terms of precision.

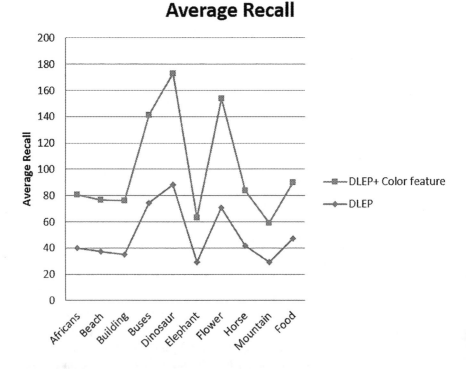

**FIGURE 7.22**   Category-wise performance in terms of recall.

provided in V. N. Gudivada and Raghavan (1995), A. W. M. Smeulders et al. (2000), H. Muller et al. (2001), R. Veltkamp and Tanase (2002), M. L. Kherfi et al. (2004).

There exists any one feature descriptor or a combination of features. Swain and Ballard (1991) devised color histogram and histogram intersection method to calculate distance between histograms. Pass et al. (1997) used CCV to create a coherent or incoherent histogram bin. To represent color distribution as well as spatial correlation, Huang et al. (1997) introduced a color feature correlogram.

Texture is an indispensable part of CBIR. Smith and Chang (1996) computed Mean and Variance of wavelet coefficients. Concept of ridgelet transform was devised by Gonde et al. (2010). S. Murala et al. (2012a) introduced the extrema concept to create a feature descriptor. L. K. Rao et al. (2015) proposed the quantized extrema patterns for image retrieval. L. K. Rao et al. (2013, 2015, 2018) introduced many feature descriptors for image retrieval. All of them are based on the relationship among pixels.

### 7.4.2 RELEVANT WORK

Ojala (2002) devised LBP for texture analysis and classification. Guo et al. (2010) introduced complete LBP scheme for classification. Considering LBP as a point edge, Reddy and Reddy proposed the LEBP (2014). Pietikainen et al. (2000), used the sign information to create feature vector.

#### 7.4.2.1 Prime Contributions

Motivated by the pattern described in Pietikainen et al. (2000), we propose a robust and effective feature descriptor. The proposed LEMP extracts more discriminating information.

1. Combination of LEBP and MLEBP is proposed for retrieval system.
2. Retrieval efficiency of the proposed method is tested on Corel image database and precision, recall and average retrieval rate (ARR) are evaluated.

The chapter is organized as follows: A brief review of CBIR is provided in Section 7.4.1. Section 7.4.2 covers a review of local patterns and line magnitude binary patterns. Section 7.4.3 explains framework, similarity measurement and results of experiments. Derived observations and conclusions are mentioned in Section 7.4.4.

### 7.4.3 RELATED LOCAL PATTERNS

#### 7.4.3.1 Local Binary Patterns

Ojala et al. (2002) proposed LBP for image texture classification. LBP is illustrated in Figure 7.41. Derivation of local pattern for a specified region is done as per the Equations (7.1)–(7.2) (Figure 7.23).

$$LBP_{P,R} = \sum_{i=0}^{P-1} 2^i \times f_1\left(h_i - h_c\right) \qquad (7.34)$$

**Example**

| 4 | 10 | 3 |
|---|----|---|
| 9 | 6 | 7 |
| 1 | 2 | 5 |

**Binary pattern**

| 0 | 1 | 0 |
|---|---|---|
| 1 |   | 1 |
| 0 | 0 | 0 |

**Weights**

| 8 | 4 | 2 |
|----|----|-----|
| 16 |   | 1 |
| 32 | 64 | 128 |

**LBP Value**

|   |    |   |
|---|----|---|
|   | 21 |   |
|   |    |   |

**LBP=1+4+16=21**

**FIGURE 7.23**  LBP computation for a 3×3 pattern.

$$f_1(n) = \begin{cases} 1 & n \geq 0 \\ 0 & else \end{cases} \tag{7.35}$$

where $h_c$ is gray level of center pixel, $h_i$ is gray value of neighbor, $P$ denotes no. of neighbors and $R$ denotes radius of neighborhood.

### 7.4.3.2  Line Edge Binary Patterns

As mentioned in P. V. N. Reddy and Prasad (2012), line edges of a center pixel are calculated with the help of eight window functions ($WF_\beta$) as follows:

$$WF_\beta = \begin{bmatrix} r & s & t \\ u & v & w \\ x & y & z \end{bmatrix}, \ \beta = 0°, 45°, \ldots, 315° \tag{7.36}$$

Values of *a to i* in Equation (7.36) are given below:

**TABLE 7.6**

**Values of *a to i***

| β | z | Y | x | w | v | u | T | s | r |
|------|----|----|----|----|----|----|----|----|----|
| 0° | 1 | −1 | 0 | 1 | −1 | 0 | 1 | −1 | 0 |
| 45° | −1 | 0 | 0 | 1 | −1 | 0 | 1 | 1 | −1 |
| 90° | 0 | 0 | 0 | −1 | −1 | −1 | 1 | 1 | 1 |
| 135° | 0 | 0 | −1 | 0 | −1 | 1 | −1 | 1 | 1 |
| 180° | 0 | −1 | 1 | 0 | −1 | 1 | 0 | −1 | 1 |
| 225° | −1 | 1 | 1 | 0 | −1 | 1 | 0 | 0 | −1 |
| 270° | 1 | 1 | 1 | −1 | −1 | −1 | 0 | 0 | 0 |
| 315° | 1 | 1 | −1 | 1 | −1 | 0 | −1 | 0 | 0 |

Further, convolution of a pattern and window function results in directional line edges (DLE) are calculated by convolution of window function with the gray values pattern.

$$DLE(\beta) = PTRN(g_c) * WF_\beta; \quad \beta = 0°, 45°, \ldots, 315° \tag{7.37}$$

$PTRN(g_c)$ denotes a 3×3 pattern of center pixel $g_c$, * denotes a convolution. LEBP (Line Edge Binary Patterns) is computed according to equation below.

$$LEBP = \sum_p 2^{(p/45)} \times f_1\left(LE(p)\right); p = 0°, 45°, \ldots, 315° \tag{7.38}$$

Once the patterns for all pixels are obtained, a histogram is built which thus represents an image.

$$H(u) = \sum_{s=1}^{P_1}\sum_{t=1}^{P_2} f_2\left(LEBP_{P,R}^{u2}(s,t), u\right); u \in [0, 255] \tag{7.39}$$

$$f_2(q,r) = \begin{cases} 1 & q = r \\ 0 & else \end{cases} \tag{7.40}$$

where $P_1 X P_2$ denotes the size.

### 7.4.3.3  Line Edge Magnitude Patterns

Line Edge Magnitude Patterns (LEMP) are extracted from line edges. Line edge computation is done according to Equation (7.37).$\beta$

In step-1, modulus of $LE$ is calculated as follows:

$$Magn(a,\beta) = |LNE(\beta)|, a = 1, 2, \ldots, (P_1 - 2) \times (P_2 - 2) \tag{7.41}$$

In the next step, obtain mean of all $Magn$ as specified in Equation (7.42).

$$M_n = mean2(Magn) \tag{7.42}$$

*LEMP is calculated according to Equation (7.43).*

$$LEMP(a) = \sum_\theta 2^{\beta/45} * h_3\left(Magn(a,\beta)\right); \beta = 0°, 45°, \ldots 315° \tag{7.43}$$

$$h_3(b) = \begin{cases} 1 & b \geq M_n \\ 0 & else \end{cases} \tag{7.44}$$

After LEMP calculation for each pixel, entire image is represented by means of a histogram.

$$Histo(q) = \sum_{z=1}^{(B_1-2)\times(B_2-2)} f_2\left(MLEP_{P,R}^{u2}(z), q\right); q \in [0, 255] \tag{7.45}$$

Here, B1 × B2 is size of input image.

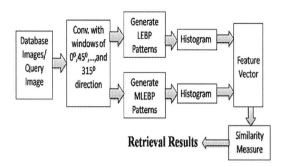

**FIGURE 7.24**   Flow diagram of the proposed system.

### 7.4.4  THE PROPOSED FRAMEWORK

In this chapter, a new method termed as magnitude line edge magnitude binary patterns (LEMP) is proposed. Further, it is integrated with LEBP to achieve improved performance. Flowchart of the proposed method is provided in Figure 7.24 and the algorithm below:

---

**ALGORITHM**

1. Get the input and convert into gray image.
2. Compute line edges in 0° to 315° directions using Equation (7.37).
3. Derive the LEMP using Equation (7.43) and build histogram using Equation (7.45).
4. Concatenate both LEBP and LEMP histograms.
5. Identify the similar images using Equation (7.46).
6. Retrieve top matches.

---

#### 7.4.4.1  Similarity Measurement

In this work, $d_l^2$ distance metric is used for calculation.

$$Dist\left(Qu,D_1\right) = \sum_{i=1}^{Vg} \left| \frac{f_{D1,i} - f_{Qu,i}}{1 + f_{D1,i} + f_{Q,i}} \right|^2 \tag{7.46}$$

where $Qu$ is image query, $Vg$ is vector length, $I_1$ is database image; $f_{D1,i}$ is $i^{th}$ feature of image D1 in database, $f_{Qu,i}$ is $i^{th}$ feature of $Qu$.

#### 7.4.4.2  Performance Evaluation and Discussions

Retrieval performance of the proposed method is tested on Corel database (1000 (DB1) and 5000 (DB2)). A comparison is made among the LBP, LEBP and the proposed method.

Precision, Recall and ARR are calculated as per Equations (7.47)–(7.49).

$$Precision\left(\text{Pr}\right) = \frac{Number\ of\ relevant\ images\ retrieved}{Total\ images\ retrieved} \tag{7.47}$$

$$Recall\left(Re\right) = \frac{Number\ of\ relevant\ images\ retrieved}{Total\ relevant\ images} \tag{7.48}$$

$$Average\ Retrieval\ Rate\left(ARR\right) = \frac{1}{X_1}\sum_{j=1}^{r_1} GRe \tag{7.49}$$

Here, $X_1$ is the number of categories.

### 7.4.4.3 Corel-1000 Database

In this chapter, we use a subset of Corel database [25], consisting of different categories. In the first phase, 1000 images in group of 100 in 10 different categories ($X_1 = 10$) are used for evaluation. Tables 7.7 and 7.8 depict retrieval results from three

**TABLE 7.7**

**Results of All Techniques, Precision on DB1 Database**

| Class | PM | LEBP | LBP |
|---|---|---|---|
| Beaches | 58.1 | 57.1 | 54.8 |
| Buildings | 75.0 | 73.8 | 63.8 |
| Buses | 98.7 | 98.1 | 95.5 |
| Dinosaurs | 98.6 | 97.9 | 98.3 |
| Elephants | 57.5 | 55.1 | 45.8 |
| Flowers | 92.0 | 92.1 | 91.8 |
| Horses | 81.0 | 80.8 | 74.3 |
| Mountains | 44.4 | 43.6 | 47.6 |
| Food | 87.7 | 86.7 | 80.9 |
| Africans | 61.1 | 58.7 | 59.5 |
| Total | **75.41** | **74.39** | **71.23** |

**TABLE 7.8**

**Recall Values on DB1**

| Class | PM | LEBP | LBP |
|---|---|---|---|
| Beaches | 38.97 | 38.87 | 32.95 |
| Buildings | 38.56 | 37.7 | 33.67 |
| Buses | 73.3 | 73.36 | 57.65 |
| Dinosaurs | 87.67 | 86.05 | 83.59 |
| Elephants | 31.15 | 30.29 | 27.32 |
| Flowers | 73.65 | 73.04 | 72.96 |
| Horses | 44.88 | 45.6 | 42.57 |
| Mountains | 29.97 | 30.01 | 27.56 |
| Food | 56.97 | 56.46 | 46.02 |
| Africans | 35.95 | 35.76 | 32.77 |
| Total | **51.17** | **50.71** | **45.71** |

methods. From Table 7.7, it is noticed that the precision of the proposed method (75.41%) is better than the other two methods. From Table 7.8, it is evident that the proposed method outperforms the other two approaches. Figure 7.25 provides a graphical comparison of all methods.

### 7.4.4.4  Corel 5000 Database (DB2)

In the second phase, a subset of Corel database is used for evaluation purpose. A total of 5000 images (DB2) with 50 categories are considered ($X_1 = 50$). Category-wise precision and recall values are shown in Figure 7.26, while average precision and recall are provided in Figure 7.27. From the two figures, it is evident that the proposed method outperforms the other methods. Query results are shown in

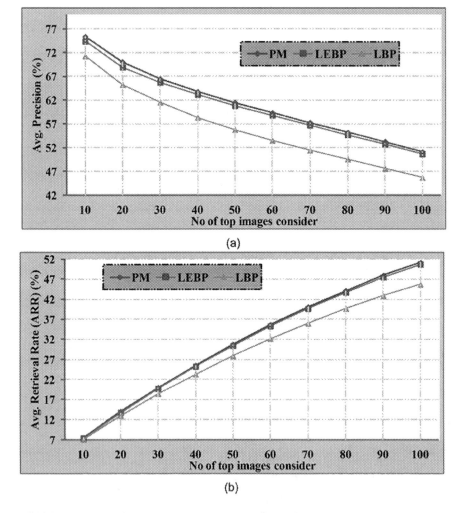

(a)

(b)

**FIGURE 7.25**  Graphical comparison of the proposed method with other methods: (a) average precision; (b) average retrieval rate (ARR) on DB1.

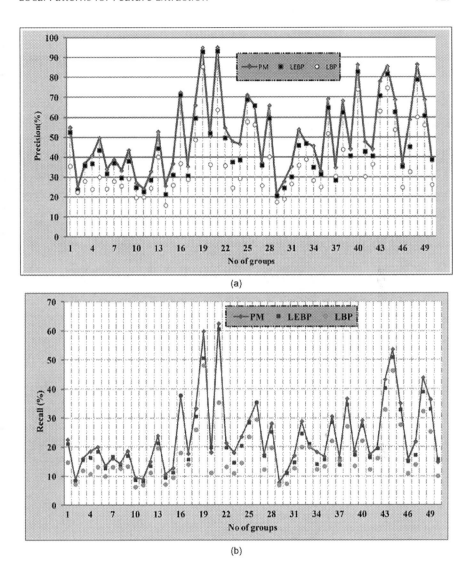

**FIGURE 7.26** Graphical comparison of the proposed method with other methods: (a) category-wise precision and (b) category-wise recall on DB2.

Figure 7.28. In all these retrieval results, top-left image is query and the remaining are similar images.

### 7.4.5 Conclusion

A novel feature descriptor named Line Magnitude Edge Pattern (LEMP) is proposed for image retrieval. Combination of LEBP and LEMP is also introduced. Performance is tested on Corel-1000 and Corel-5000 databases. Experimental results indicate that the LEMP method outperforms other methods. There exists 2–3% of improvement in precision and recall values.

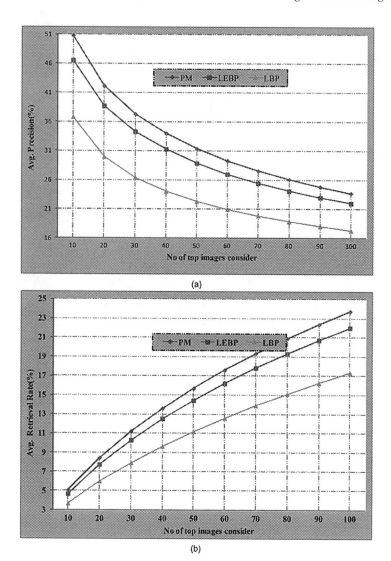

**FIGURE 7.27** Comparison of the proposed method with other methods: (a) average precision and (b) average retrieval rate (ARR) on DB2.

## 7.5 MULTIPLE COLOR CHANNEL LOCAL EXTREMA PATTERNS FOR IMAGE RETRIEVAL

### 7.5.1 INTRODUCTION

CBIR describes an image according to some explored features. It resolves the key issues in traditional annotation method by extracting the features from visual data of the image such as texture, shape and color. Some contributions made for the improvement in the efficacy of retrieval systems are discussed in M. L. Kherfi et al. (2004),

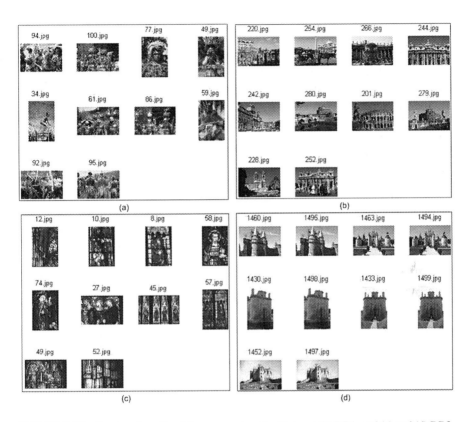

**FIGURE 7.28**    Retrieved results of the proposed method (a) and (b) DB1; and (c) and (d) DB2.

R. Dutta et al. (2008). Texture is one among the primary features used in the automated retrieval system such as CBIR. Many researchers proposed several retrieval methods by using the texture as the key element V. N. Gudivada and Raghavan (1995), A. W. M. Smeulders et al. (2000) H. Muller et al. (2001), R. Veltkamp and Tanase (2002), M. L. Kherfi et al. (2004).

A feature vector, LBP Ojala et al. (2002), became the foundation for a few retrieval algorithms. Takala et al. (2005) introduced block-based LBP. Yao and Chen (2003) proposed LEPSEG, LEPINV for texture description. S. Murala et al. introduced a novel frame work, DLEP for retrieval (2012b). K. Rao et al. (2016) introduced LQMeP for image retrieval. L. K. Rao et al. (2016) proposed local color oppugnant quantized patterns. L. K. Rao and Venkata Rao (2015) have proposed the local quantized extrema patterns.

In the present work, the texture from the image is derived by using a new feature descriptor. Performance is evaluated on Corel image repository.

Contents are arranged as follows: Section 7.5.1 covers a review of retrieval approaches. Two primary ways of extracting the information are given in Section 7.5.2. Section 7.5.3 provides a detailed framework and explains the algorithm of the proposed MCLEP method. In Section 7.5.4, summary of results is given. Conclusion is provided in Section 7.5.5.

## 7.5.2 Relevant Work

### 7.5.2.1 Local Quantized Extrema Patterns

By applying the quantization to DLEP, L. K. Rao et al. (2015) proposed a feature descriptor named Local Quantized Extrema Patterns (LQEP). In the first step, geometric structures are separated from an image in specific directions. Extremal operation is performed on the derived structures. Figure 7.3 shows the calculation of the pattern described in L. K. Rao et al. (2015). For instance, computation of pattern in horizontal direction is done using the equation. Then the four directional extremas (DE) in $0°$, $45°$, $90°$ and $135°$ directions are obtained as follows (Figures 7.29).

$$DE(I(g_c))\big|_{0°}$$
$$= \left\{ y_2((g_1),I(g_4),I(g_p)); y_2(I(g_2),I(g_5),I(g_p)); y_2(I(g_3),I(g_6),I(g_p)) \right\} \quad (7.50)$$

Here,

$$y_2(j,k,p) = \begin{cases} 1 & if(j > p)or(k > p) \\ 1 & if(j < p)or(k < p) \\ 0 & otherwise \end{cases} \quad (7.51)$$

LQEP is computed according to Equation (7.52).

$$LQEP = \left[ DE\big(I(g_c)\big)\big|_{0°}, DE\big(I(g_c)\big)\big|_{45°}, DE\big(I(g_c)\big)\big|_{90°}, DE\big(I(g_c)\big)\big|_{135°} \right] \quad (7.52)$$

At the end, image is converted to LQEP maps with range of pixel intensities from 0 to 4095. A histogram is built using Equation (7.53).

$$Hstg_{LQEP}(s) = \sum_{b=1}^{Q_1} \sum_{c=1}^{Q_2} y_2\big(LQEP(b,c),s\big); s \in [0,4095]; \quad (7.53)$$

## 7.5.3 The Proposed Method

The proposed MCLEP inherits the salient properties of LQEP and LBP, i.e., the relative difference of information between a central pixel and its neighbors in a local neighborhood. MCLEP is capable of capturing the texture information at various places for any color channels. Let $M^x$ be the $x^{th}$ channel of an image M with X,Y,C

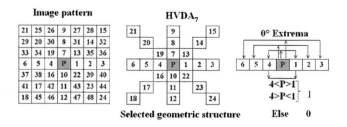

**FIGURE 7.29**   LQEP calculation for a given 7×7 pattern.

as size. Here, $x \in [1,N]$ and N denotes the total number of color channels. Let the N neighbors at the radius D of any pixel with location co-ordinates $J'(m,n)$ for $m \in [1,P]$ and $n \in [1,Q]$ be defined as $J_t^U (x,y)_{\text{where}} s \in [1,R]$.

$$MCLEP_{N,D}^{t,d} (x,y) = \left\{ MCLEP^1 (x,y); MCLEP^2 (x,y) \right\} \qquad (7.54)$$

$$MCLEP^2 (x,y) = LQEP_{N,D}^{t,d} (x,y) \qquad (7.55)$$

$$MCLEP^1 (x,y) = LBP_{N,D}' (x,y) \qquad (7.56)$$

Here, d=1, 2, 3, 4 represents various distances at which the MCLEP is computed. LBP and LQEP are computed using the equations above in (7.1)–(7.5). MCLEP for $t^{th}$ space contains one LBP and four LQEPs.

### 7.5.4  A MCLEP FEATURE VECTOR

Two major components of pattern recognition applications are feature representation and matching/classification techniques. For delivering competitive performance, it is essential to ensure robust implementation of both parts. MCLEP histogram HSTG is generated as the feature vector for robust image feature representation, where HSTG ={HSTG$_x$}, $x \in [1,N]$. Let $J(m, n)$ be a color image of size $X,Y,C$ with $W$ intensity levels, i.e., $I \in \{0,1,...,W-1\}$, where $(m, n)$ is the location coordinate for $m \in [1, X]$ and $n \in [1,Y]$. The MCLEP histogram $HSTG(.)$ is computed as given in Equation (7.57):

$$HSTG_t (MDLP) = \sum_{i=1}^{N_1} \sum_{j=1}^{N_2} f \left( MCLEP_{Nb,R}^{t,d} (x,y), l \right) \qquad (7.57)$$

$$f(x,y) = \begin{cases} 1 & x = y \\ 0 & else \end{cases} \qquad (7.58)$$

## ALGORITHM

*I/P: IMAGE; O/P: RETRIEVED IMAGE*

1. Load the image.
2. Compute the MCLEP.
3. Match the query image with the database image.
4. Retrieve the similar images.

### Query Matching

A distance metric such as given in Equation (7.59) is used to compute the distance between the images under consideration. Here, $s_R = (s_{R1}, s_{R2}, \ldots \ldots s_{RLg})$ denotes the feature vector of query image. Similarity matching is carried out using $d_l$ distance metric as mentioned in Equation (7.12). Each image is denoted by feature vectors $_{DBSj} = (s_{DBSj1}, s_{DBSj2}, \ldots \ldots s_{DBSjLg}); j = 1, 2, \ldots \ldots, |DBS|$.

$$d_1 distance: d\left(R, I_1\right) = \sum_{k=1}^{Lg} \left| \frac{s_{DBS_{jk}} - s_{R,k}}{1 + s_{DBS_{jk}} + s_{R,k}} \right| \qquad (7.59)$$

where $s_{DBS_{jk}}$ represents $k^{th}$ feature of $j^{th}$ image of database $|DBS|$.

### 7.5.5  EXPERIMENTAL RESULTS AND DISCUSSIONS

Efficiency of the devised method is calculated by means of the experiments on standard databases. Corel image repository is used for this purpose. In every experiment, each database image becomes a query image. Accordingly, the proposed framework identifies $n$ images $X=(x_1, x_2, ...,x_n)$ as per the shortest distance. Performance is evaluated in terms of average precision/retrieval precision (ARP), average recall/retrieval rate (ARR) as indicated below: Evaluation metrics for a query image $I_q$ are calculated according to the equation given below.

$$Prec: P\left(I_q\right) = \frac{Number\ of\ related\ images\ retrieved}{Total\ images\ retrieved} \qquad (7.60)$$

$$Avg.\ Retrieval\ Prec: ARP = \frac{1}{|DBS|} \left| \sum_{k=1}^{|DBS|} P\left(I_k\right) \right| \qquad (7.61)$$

$$Recall: Re\left(I_q\right) = \frac{Number\ of\ related\ images\ retrieved}{Total\ relevant\ images\ in\ database} \qquad (7.62)$$

$$Avg.Retreival\ Rate: ARR = \frac{1}{|DBS|} \left| \sum_{k=1}^{|DBS|} Re\left(I_k\right) \right| \qquad (7.63)$$

### TABLE 7.9
### Results of Various Methods on Corel Database

| Image Repository | Evaluation Metric | CS_LBP | LEP SEG | LEP INV | BLK_LBP | LBP | DLEP | LQEP | MCLEP |
|---|---|---|---|---|---|---|---|---|---|
| | %Precision | 25.4 | 26.2 | 27.6 | 29.7 | 22.2 | 30.1 | 31.9 | 33.5 |
| Corel-10k | %Recall | 18.0 | 19.6 | 20.8 | 21.6 | 19.7 | 23.8 | 24.9 | 25.6 |

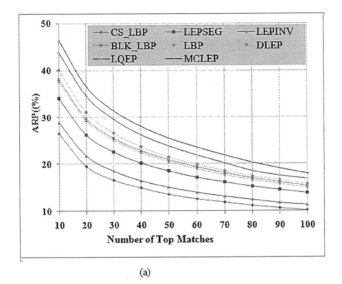

(a)

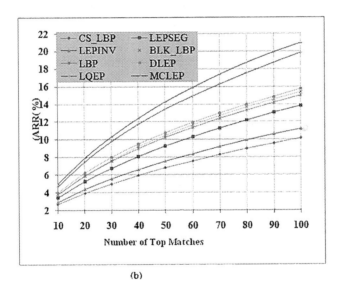

(b)

**FIGURE 7.30**   Comparison of various methods in terms of (a) ARP (b) ARR.

### 7.5.5.1   Corel-10k

Corel-10k is a collection of 10000 images classified into different categories. Results of precision and recall are provided in the Table 7.9. Comparisons of the results are shown in Figure 7.30 whereas the retrieved images are provided in Figure 7.31. The improvement in performance is mainly due to the discriminating

**FIGURE 7.31**   Retrieved images using the proposed method.

ability of the proposed method against any other method mentioned in Table 7.9 (Figures 7.30 and 7.31).

### 7.5.5.2   ImageNet-25K

ImageNet-25K is a composition of 25000 images. These are classified into different categories. Performance of MLCEP is determined by means of ARR and ARP (Figures 7.32).

### 7.5.6   Conclusion

A new method for image retrieval is introduced in this chapter. It explores the directional data using multiple color channel local extrema pattern from color planes. Two types of repositories are used to test the effectiveness of the method. Results after investigation exhibit a substantial improvement in the performance metrics such as precision and recall.

By extracting the facial properties of any image, the proposed approach can be extended in the field of face recognition.

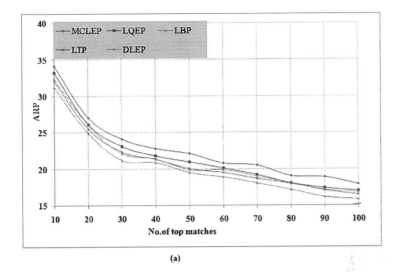

(a)

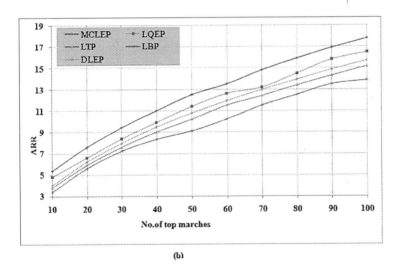

(b)

**FIGURE 7.32** Comparison of various methods in terms of (a) ARP (b) ARR.

## 7.6 INTEGRATION OF MDLEP AND GABOR FUNCTION AS A FEATURE VECTOR FOR IMAGE RETRIEVAL SYSTEM

### 7.6.1 INTRODUCTION

Classification and segmentation became very critical in analyzing the texture of an image. In R. M. Haralick et al. (1973a), usage of texture feature for the classification of images was mentioned. Arivazhagan and Ganesan (2003) proposed an approach to classify the texture that uses wavelet transform. Wavelet packet frames and Gaussian mixture model was specified in S. C. Kim and Kang (2007) to classify the texture and segmentation. Gabor wavelets played an important role in texture classification for rotation invariant features Arivazhagan and Ganesan (2006).

The existing Magnitude Directional Local Extrema Pattern (MDLEP) extracts the directional and magnitude information of edges based on the minima or maxima in vertical, horizontal, diagonal, anti-diagonal directions of an image. In this chapter, we propose a new method which combines the color feature and MDLEP in order to improve the performance of the existing MDLEP. This chapter is organized as follows. Introduction and related work is discussed in Section 7.6.1. Different types of local patterns are reviewed in Section 7.6.2. Section 7.6.3 mentions the proposed work for image retrieval system. Section 7.6.4 contains the results and the conclusions are given in Section 7.6.5.

### 7.6.1.1 Related Work

Subrahmanyam et al. (2012) proposed an approach called DLEP as a feature vector for texture analysis and classification. Reddy and Reddy (2014) extended the DLEP by taking the magnitude into consideration. The MDLEP is different from the existing LBP and other extensions in terms of directional information.

### 7.6.2 LOCAL PATTERNS AND VARIATIONS

### 7.6.2.1 Directional Local Extrema Patterns

The concept of LBP was used by Subrahmanyam et al. (2012) to design a new feature descriptor called Directional Local Extrema Patterns (DLEP). In this method, two neighboring pixel intensities in a direction are compared with the pixel at the center position to assign either 0 or 1. In other words, it describes the spatial structure of the local texture based on the local extrema of center gray pixel $x_c$. The maxima and minima values in four directions are obtained by taking the difference between the center pixel and all its neighbors.

The calculation is as mentioned below.

$$I'(x_i) = I(x_c) - I(x_i); i = 1, 2, \ldots 8 \tag{7.64}$$

The local extremas are based on the equations given below.

$$\hat{I}_\beta(xc) = Y_3\left(I'(x_i) * I'(x_{j+4})\right); j = (1 + \beta/45) \tag{7.65}$$
$$\forall_\beta = 0°, 45°, 90°, 135°$$

$$f_3\left(I'(g_j), I'(g_{j+4})\right) = \begin{cases} 1 & I'(g_j) \, \mathrm{XI}'(g_{j+4}) \geq 0 \\ 0 & \text{else} \end{cases}$$

$$Y_3\left(I'(x_j), I'(x_{j+4})\right) = \begin{cases} 1 & I'(x_j) * I'(x_{j+4}) \geq 0 \\ 0 & \text{else} \end{cases} \tag{7.66}$$

The DLEP is computed as ($\beta=0°, 45°, 90°$ and $135°$) follows:

$$DLEP\left(I\left(x_c\right)\right)\big|\beta = \{\hat{I}_\beta\left(x_c\right); \hat{I}_\beta\left(x_i\right); \hat{I}_\beta\left(x_z\right); \ldots \hat{I}_\beta\left(x_8\right) \tag{7.67}$$

The details of DLEP can be found in Figure 7.12. In the next step, the given image is converted into DLEP images with values ranging from 0 to 511.

After calculation of DLEP, the whole image is represented by building a histogram based on the equation mentioned below.

$$H_{DLEP|\beta}\left(l\right) = \sum_{m=1}^{Z_1}\sum_{n=1}^{Z_2} Y_2\left(DLEP\left(m,n\right)\big|_\alpha, \ell\right); \tag{7.68}$$

$$\ell \in \left[0, 511\right]$$

where the size of input image is $Z_1 \cdot Z_2$. The procedure for calculation of DLEP for center pixel marked in blue color is presented in Figure 7.13. The directions are evaluated using the local difference between the center pixel and its neighbors.

As an example, the DLEP in $90°$ direction for a center pixel marked in blue color is shown in Figure 7.12. For the center pixel value 36, it can be observed that two neighboring pixels are leaving. Therefore, this pattern is represented as 1. In the same way the rest of the bits of DLEP pattern are calculated and the result is 110011110. In the same way, the DLEPs are computed in $0°, 450$ and $135°$ directions.

### 7.6.2.2 Magnitude Directional Local Extrema Patterns (MDLEP)

Reddy et al. [13] proposed a method to increase the performance by considering the magnitudes of the local patterns. The magnitude patterns are calculated as specified in the equation below.

$$\hat{I}_{M\beta}\left(xc\right) = Y_3\left(I'\left(x_i\right)*I'\left(x_{j+4}\right)\right); j = \left(1+\beta/45\right) \tag{7.69}$$

$$\forall_\beta = 0°, 45°, 90°, 135°$$

$$Y_4\left(I'\left(x_j\right), I'\left(x_{j+4}\right)\right) = \begin{cases} 1 & abs\left(I'\left(x_j\right)\right) + abs\left(I'\left(x_{j+4}\right)\right) \geq \text{Thrs} \\ 0 & else \end{cases} \tag{7.70}$$

$$\text{Thrs} = \frac{1}{Z_1 \times Z_2}\sum_{b=1}^{z_1}\sum_{c=1}^{z_2}\left(abs\left(I'\left(x_j\right)\big|_{(b,c)}\right) + abs\left(I'\left(x_{j+4}\right)\big|_{(b,c)}\right)\right) \tag{7.71}$$

The MDLEP in $0°, 45°, 90°$ and $135°$ directions is defined as

$$MDLEP\left(I\left(x_c\right)\right)\big|\beta = \left\{\hat{I}_{M\beta}\left(x_c\right); \hat{I}_{M\beta}\left(x_1\right); \hat{I}_{M\beta}\left(x_2\right); \ldots \hat{I}_{M\beta}\left(x_8\right)\right\} \tag{7.72}$$

Subsequent to the MDLEP calculation, the whole image is represented by a histogram as per Equation (7.72) Figures 7.33–7.35.

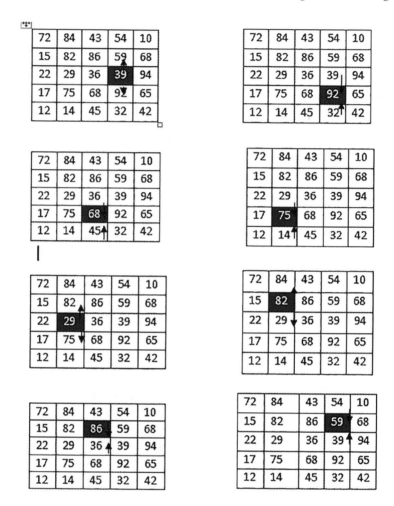

**FIGURE 7.33**  Illustration of DLEP for 5 × 5 pattern.

|           | $0_{(27)}$ | $1_{(29)}$ | $2_{(80)}$ | $3_{(87)}$ | $4_{(88)}$ | $5_{(13)}$ | $6_{(78)}$ | $7_{(85)}$ | $8_{(63)}$ | DLEP |
|-----------|------|------|------|------|------|------|------|------|------|------|
| P (0°)    | 0 | 0 | 0 | 0 | 1 | 1 | 0 | 1 | 0 | 26 |
| Q(45°)    | 1 | 0 | 0 | 1 | 1 | 1 | 1 | 0 | 1 | 317 |
| R(90°)    | 1 | 1 | 0 | 0 | 1 | 1 | 1 | 1 | 0 | 415 |
| S(135°)   | 1 | 1 | 0 | 0 | 0 | 1 | 1 | 1 | 0 | 398 |

**FIGURE 7.34**  Example to compute DLEP in 90° direction (110011110).

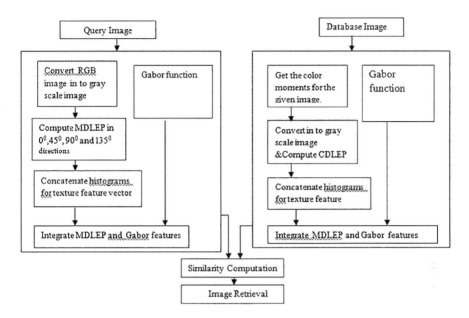

**FIGURE 7.35**   The proposed GMDLEP for retrieval system.

### 7.6.3   PROPOSED **CMDLEP** SYSTEM

**THE PROPOSED ALGORITHM**

1. Apply the Gabor function on the query image.
2. Compute the local extrema in $0°, 45°, 90°$ and $135°$ directions.
3. Compute the MDLEP patterns in four directions mentioned in step 2.
4. Get a histogram for the MDLEP patterns obtained in step 3 and concatenate to get the texture feature vector.
5. Combine these two features to get a feature vector that can be used in the process of image retrieval.

*Query Matching*

Once the features are extracted, the feature vector for query image is obtained. In the same manner, feature vectors for all images in the database are also calculated. To select the more relevant image to the query image, the distance between query image and database images is calculated.

### 7.6.4   EXPERIMENTAL RESULTS

Performance of the proposed method is evaluated on standard corel-1k database [15]. The precision (Pr) and recall (Re) values are computed as per the relationship mentioned here under (Figures 7.36 and 7.37).

**TABLE 7.10**

**The Top Ten Results for Different Categories of the Database are Shown Below**

| Category | Existing MDLEP | MDLEP+ Color Feature | Category | MDLEP | MDLEP+ Color Feature |
|---|---|---|---|---|---|
| Africans | 61.3 | 64.4 | Africans | 39.25 | 43.7 |
| Beach | 51.25 | 53.7 | Beach | 33.82 | 37.5 |
| Building | 57.85 | 62.6 | Building | 31.96 | 36.6 |
| Buses | 94.4 | 98.4 | Buses | 73.57 | 77.9 |
| Dinosaur | 97.85 | 99.1 | Dinosaur | 90.28 | 94.5 |
| Elephant | 48.9 | 64.8 | Elephant | 30.53 | 34.7 |
| Flower | 89.1 | 93.5 | Flower | 69.32 | 77.8 |
| Horse | 66.2 | 79.4 | Horse | 36.16 | 45.4 |
| Mountain | 39.4 | 48.5 | Mountain | 29.35 | 34.1 |
| Food | 75.35 | 90 | Food | 45.3 | 43.5 |
| Average Precision (%) | 68.16 | 75.44 | Average Recall (%) | 47.954 | 52.57 |

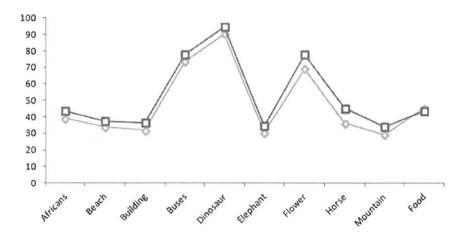

**FIGURE 7.36**   Average precision of MDLEP (green) and CMDLEP (red).

$$Pr = \frac{No.\,of\ images\ retrieved\ those\ are\ relevant}{No.\,of\ relevant\ images\ retrieved}$$

$$Re = \frac{No.\,of\ relevant\ images\ retrieved}{No.\,of\ relevant\ images\ in\ the\ database}$$

### 7.6.5   Conclusion

It is proved that the proposed method is outperforming the existing MDLEP both in terms of precision and recall. The combination of Gabor and MDLEP is exploring more information present in the image when compared to LBP and other related local pattern operators.

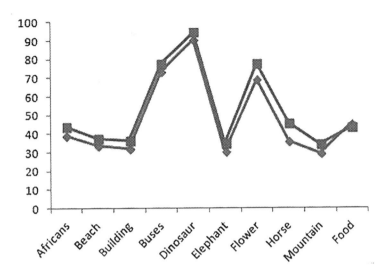

**FIGURE 7.37**    Average recall of MDLEP (red) and CMDLEP (blue).

## 7.7    CONTENT-BASED MEDICAL IMAGE RETRIEVAL USING LOCAL CO-OCCURRENCE PATTERNS

### 7.7.1    INTRODUCTION

CBIR is also useful in biomedical image retrieval to identify the location of the disease for decision support. Different methods for medical image retrieval can be found in H. Muller et al. (2004), S-N Yu et al. (2005), H. Pourghassem and Ghassemian (2008). Extensive literature survey on different features for CBIR is presented in M. L. Kherfi et al. (2004), A. W. M. Smeulders et al. (2000), W. P. Santos et al. (2009) M. M. Rahman and Bhattacharya (2010) and R. Joe Stanleya et al. (2011).

The classification of praxis using objective dialectical method (ODM) has been proposed in W-J. Wu et al. (2012), T. Ojala et al. (1996). Various methods on biomedical image retrieval are given in T. Ahonen et al. (2006), Zhang et al. (2010a).

The methods using Gabor transform (GT), co-occurrence matrix, etc., are computationally more expensive. To address this complexity issue, The LBP operator was introduced by Ojala et al. (2002) for texture classification. The LBP was used extensively by many researchers to retrieve biomedical images and in other applications L. Sorensen et al. (2010), D. S. Marcus et al. (2007). The LBP was enhanced to get nth order derivative called Local Directional Pattern (LDP) B. Zhang et al. (2010a). To overcome the limitations of LBP and LDP, local ternary pattern (LTP) X. Tan and Triggs (2010) has been introduced for face recognition. The idea of local features LBP, LTP and LDP has motivated us to propose the LCoP for biomedical image retrieval. In the proposed local co-occurrence patterns (LCoP) occurrence of two similar edges that is computed by first-order derivatives at possible directions in a circular neighborhood is encoded, unlike LDP which encodes the occurrence of two different edges that are calculated by $1^{st}$-order derivatives in a particular direction.

Further, the performance is improved by combining it with the joint histogram between the gray-scale value of center pixel. The effectiveness of the proposed method is confirmed by combining it with GT. The performance of the proposed method is tested in terms of average retrieval precision (ARP) on a biomedical image database.

### 7.7.2 LOCAL PATTERNS

#### 7.7.2.1 Local Binary Patterns

The LBP operator considers eight surrounding pixels, taking the center pixel gray value as a threshold. A binary "1" or "0"is generated based on threshold value. The *LBP* operator outputs w.r.t. all pixels in the image can be gathered to form a histogram. Figure 7.38 gives an example of LBP.

$$LBP_{P,R} = \sum_{i=0}^{P-1} 2^i \times f\left(g_i - g_c\right) \tag{7.73}$$

$$f\left(x\right) = \begin{cases} 1 & x \geq 0 \\ 0 & else \end{cases} \tag{7.74}$$

where
$g_c$=Center pixel gray value;
$g_i$= gray value of neighborhood pixels;
$P$= Number of neighbors;
$R$= Radius of the neighborhood.

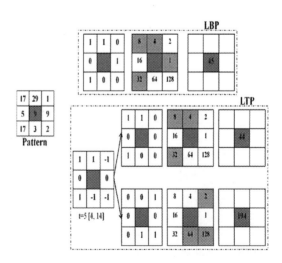

**FIGURE 7.38** Calculation of LBP and LTP operators.

### 7.7.2.2 Local Ternary Patterns

The LTP is a three-value code, proposed by X. Tan and Triggs (2010). The gray values in a region of width $\pm t$ around $g_c$ are quantized to 0, 1 and $-1$. The indicator $f(x)$ is replaced with a three-valued function as defined in Equation (7.75)

$$f\left(x,g_c,t\right)=\begin{cases}1 & x\geq g_c+t \\ 0 & \left|x-g_c\right|<t \\ -1 & x\leq g_c-t\end{cases} \quad (7.75)$$

where t = user-defined threshold value.

The procedure to calculate LTP for 3×3 patterns is given in Figure 7.38. In the example, we consider the threshold as 5, and hence the tolerance interval becomes (4, 14).

### 7.7.2.3 Local Derivative Patterns

Zhang et al. (2010b) introduced the local derivative patterns (LDPs) for face recognition. They considered LBP as a non-directional $1^{st}$-order operator and extended it to higher-orders ($n^{th}$-order) called LDP. LDP contains more detailed discriminative information when compared to LBP.

To calculate the $n^{th}$-order LDP, the $(n-1)^{th}$-order derivatives are calculated along $0°, 45°, 90°$ and $135°$ directions, denoted as $I_\alpha^{(n-1)}\left(g_c\right)\Big|_{\alpha=0°,45°,90°,135°}$. Finally $n^{th}$-order LDP is calculated as

$$LDP_\alpha^n\left(g_c\right)=\sum_{p=1}^{P}2^{(p-1)}\times f_2(I_\alpha^{(n-1)}\left(g_c\right),I_\alpha^{(n-1)}\left(g_p\right))\Big|_{P=8} \quad (7.76)$$

$$f_2\left(x,y\right)=\begin{cases}1 & if\ x.y\leq 0 \\ 0 & else\end{cases} \quad (7.77)$$

Figure 7.39 illustrates the calculation of LDP in $0°$ direction and the detailed discussion is available in B. Zhang et al. (2010a).

### 7.7.2.4 Local Co-Occurrence Patterns

The idea of LDP and LTP has motivated us to propose the LCoP for biomedical image retrieval. For a center pixel, LCoP is calculated based on the $1^{st}$-order derivatives in eight directions as shown in Figure 7.39.

The $1^{st}$-order derivatives for a given center pixel $(g_c)$ are calculated as follows:

$$\tilde{I}_{P,R}\left(g_i\right)=I_{P,R}\left(g_i\right)-I_{P,R}\left(g_c\right);\ i=1,2,\ldots,P \quad (7.78)$$

$$\tilde{I}_{P,R+1}\left(g_i\right)=I_{P,R+1}\left(g_i\right)-I_{P,R}\left(g_i\right);\ i=1,2,\ldots,P \quad (7.79)$$

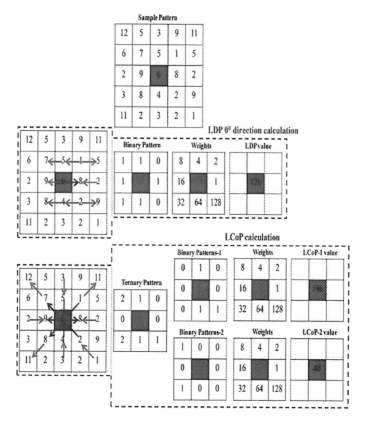

**FIGURE 7.39**   Example for LDP and LCoP calculation.

After the calculation of 1ˢᵗ-order derivatives, we encode them based on the sign of derivative as follows:

$$I^1_{P,R}(g_i) = f_1\left(\tilde{I}_{P,R}(g_i)\right) \tag{7.80}$$

$$I^1_{P,R+1}(g_i) = f_1\left(\tilde{I}_{P,R+1}(g_i)\right) \tag{7.81}$$

The LCoP is defined as:

$$LCoP = \begin{bmatrix} f_3\left(I^1_{P,R}(g_1), I^1_{P,R+1}(g_1)\right), \\ f_3\left(I^1_{P,R}(g_2), I^1_{P,R+1}(g_2)\right), \ldots \\ \ldots, f_3\left(I^1_{P,R}(g_P), I^1_{P,R+1}(g_P)\right) \end{bmatrix} \tag{7.82}$$

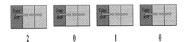

**FIGURE 7.40**    Co-occurrence calculation with respect to the center pixel.

$$f_3(x,y) = \begin{cases} 1 & \textit{if } x = y = 0 \\ 2 & \textit{if } x = y = 1 \\ 0 & \textit{else} \end{cases} \tag{7.83}$$

LCoP is a ternary pattern (0, 1, 2) which is further separated into two binary patterns by adopting the concept of LTP. The detailed representation of co-occurrence calculation for two types of edges is given in Figure 7.40.

In order to decrease the computational cost, the uniform patterns are considered D. S. Marcus et al. (2007). In this chapter, patterns those have less than or equal to two discontinuities in the circular binary representation are referred to as the uniform patterns and the remaining patterns are referred to as non-uniform. Therefore, the distinct uniform patterns for a given query image would be $M(M-1)+2$ deprived of rotational invariance.

After identifying the local pattern, *PTN* (LBP or LTP or LDP or LCoP), the whole image is represented by constructing a histogram using Equation (7.84)

$$H_S(l) = \frac{1}{N_1 \times N_2} \sum_{j=1}^{N_1} \sum_{k=1}^{N_2} f_4 \big( PTN(j,k),l \big); l \in [0, L-1] \tag{7.84}$$

$$f_4(x,y) = \begin{cases} 1 & \textit{if } x = y \\ 0 & \textit{else} \end{cases} \tag{7.85}$$

where $L$ is the number of bins in an histogram and $N_1 \times N_2$ is the size of input image.

Figure 7.41 shows two sample images which are selected from the different groups of OASIS-MRI database to analyze the different techniques in terms of histogram. Figure 7.41 shows the histograms of various approaches on two sample images (shown in Figure 7.41). From Figure 7.42, it is evident that the feature LCoP outperforms the LBP, LDP and LTP in differentiating the images.

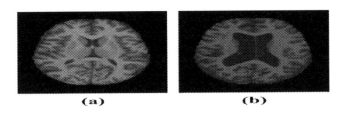

(a)                                       (b)

**FIGURE 7.41**    Sample images from OASIS-MRI database.

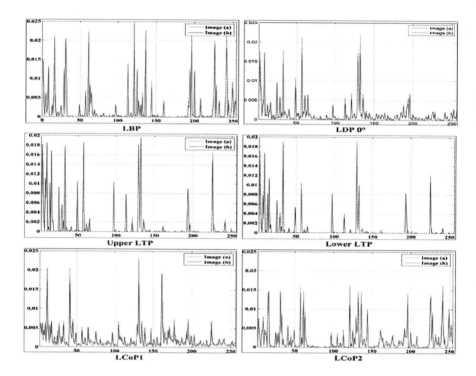

**FIGURE 7.42**   Histograms of various methods on two sample images.

### 7.7.3  FRAMEWORK OF THE PROPOSED SYSTEM

Figure 7.43 illustrates the framework of the proposed system and algorithm for the same is given below.

**ALGORITHM**

*Input: Query image*                  *Output: Retrieval results*

1. Load the query image.
2. Calculate the 1st-order derivatives at R and R+1 with respect to the center pixel.
3. Calculate the co-occurrence between the derivatives.
4. Make the LCoP which is a ternary pattern.
5. Convert LCoP into two binary patterns.
6. Construct the histograms between the gray value of center pixel and LCoPs.
7. Form the feature vector by concatenating the histograms.
8. Compare the query image with the images in the database using Equation (7.84).
9. Retrieve the images based on the best matches found.

This algorithm is also applied on GT sub-bands for LCoP with Gabor transform (GLCoP).

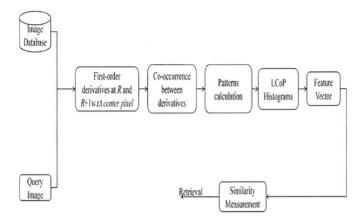

**FIGURE 7.43** Flowchart of the proposed retrieval system.

### 7.7.3.1 Similarity Measure

The following equation is used as a distance metric to obtain the similarity between the query image and database images.

$$D(Q,I_1) = \sum_{i=1}^{Lg} \left| \frac{f_{I_1,i} - f_{Q,i}}{1 + f_{I_1,i} + f_{Q,i}} \right| \tag{7.86}$$

where

Q refers to query image;

$I_1$ is image in database;

Lg is feature vector length;

$f_{I1,i}$ is $i^{th}$ feature of image $I_1$ in the database;

$f_{Q,i}$ is $i^{th}$ feature of query image Q.

### 7.7.3.2 Evaluation Measures

The ARP judge the performance of the proposed method and those are calculated by Equations (7.87)–(7.88).

For the query image $I_q$, the precision (P) and recall (R) are defined as follows:

$$Precision\ P(Iq) = \frac{Number\ of\ relevant\ images\ retrieved}{Total\ number\ of\ images\ retrieved} \tag{7.87}$$

$$ARP = \frac{1}{|DB|} \sum_{i=1}^{DB} P(I_i) \Bigg|_{n \geq 10} \tag{7.88}$$

### 7.7.4 EXPERIMENTAL RESULTS AND DISCUSSIONS

In order to analyze the performance of our algorithm for biomedical image retrieval, an experiment is conducted on Open Access Series of Imaging Studies (OASIS), a

**TABLE 7.11**

**MRI data acquisition details**

| Sequence | MP-RAGE |
|---|---|
| TR (msec) | 9.7 |
| TE (msec) | 4.0 |
| Flip angle (o) | 10 |
| TI (msec) | 20 |
| TD (msec) | 200 |
| Orientation | Sagittal |
| Thickness, gap (mm) | 1.25, 0 |
| Resolution (pixels) | 176×208 |

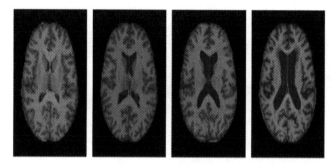

**FIGURE 7.44**    Sample images from OASIS-MRI database (one image per category).

medical image database. It consists of a cross-sectional collection of 421 subjects aged between 18 and 96 years. The MRI acquisition details are given in Table 7.11.

In our experiment, we grouped these 421 images into four categories (124, 102, 89 and 106 images) based on the shape of ventricular in the images (Figure 7.44).

From the experiment, the following inference is drawn in terms of ARP.

LCoP outperforms the LDP, LTP, LBP (without uniform two, with uniform two and rotational invariant uniform two) and DBWPu2 in terms of ARP.

Figure 7.45(a)–(c) show the retrieval performance of the proposed method and other existing methods in terms of ARP. Figure 7.45(d)–(f) illustrate the group-wise performance of various methods with and without GT in terms of Precision, ARP.

## 7.7.5 Conclusion

In this chapter, a new approach, LCoP is presented for computerized biomedical image retrieval; unlike LDP, LCoP encodes the occurrence of two similar edges which are calculated by $1^{st}$-order derivatives at all possible directions in a circular neighborhood. The performance of the LCoPs is enhanced by integrating it with the joint histogram between the gray-scale value of center pixel and its LCoP value. An experiment is conducted on a standard dataset, and the results showed a considerable improvement in terms of precision and ARP as compared to other existing methods on similar databases.

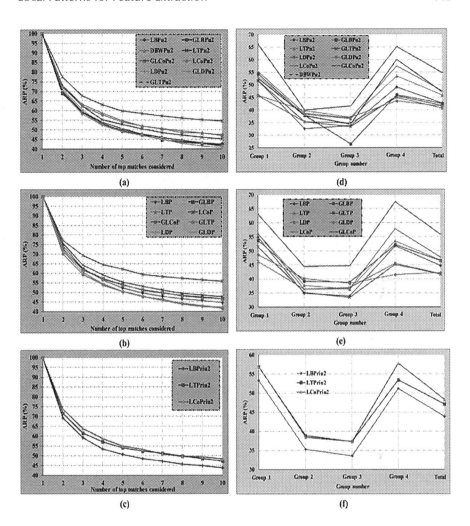

**FIGURE 7.45** (a)–(c) Comparison of the proposed method with other existing methods as function of number of top matches. (d)–(f) Group-wise performance of various methods in terms of ARP on OASIS-MRI database.

## 7.8 COLOR-BASED MULTI-DIRECTIONAL LOCAL MOTIF XoR PATTERNS FOR IMAGE RETRIEVAL

### 7.8.1 INTRODUCTION

CBIR system mainly depends on feature extraction. It is the technique utilized for extracting the features from a given image. There is no single best representation of an image for all perceptional subjectivity because the user may capture the photographs in various conditions. The extensive literature survey on CBIR is available in M. Kokare et al. (2002), A. W. M. Smeulders et al. (2000).

In recent times, local feature descriptors are gaining more popularity in the field of pattern recognition. The L. B. P. Ojala et al. (2002) feature has emerged as a silver lining in the field of texture classification and retrieval. Success of LBP variant in terms of speed and performance is reported in many research areas. Ojala et al. (2002) proposed LBPs, which are converted to a rotational invariant version for texture classification B. Zhang et al. (2010a). The different local feature descriptors are given in S. K. Vipparthi and Nagar (2014a), Vipparthi et al. (2015). The concept of DBC B. Zhang et al. (2010a), expert image retrieval system and multi-joint histogram-based motif and texton S. K. Vipparthi et al. (2014b) motivated us to propose CMDLMXoRP.

### 7.8.2 FEATURE EXTRACTION METHODS

#### 7.8.2.1 HSV Color Space and Color Quantization

In HSV color space are defined in terms of hue (H), saturation (S) and intensity value (V). These three attributes are perceived about color. The H component specifies the type of the color and it ranges from $0°$ to $360°$. The relative purity of the color is stated by S color component which ranges from 0 to 1. Image having lower saturation value will have more grayness and faded color will appear on the image. The brightness of the color is represented by V component which ranges from 0 to 1. It is well demonstrated that the color descriptors provide valuable information from an image. The proposed descriptor uses the HSV color space, particularly the H, S and V color channels are quantized.

#### 7.8.2.2 Directional Binary Code

The directional binary code (DBC) was proposed by Bao chang et al. (2010). This encodes the directional edge information. Given an image I, its first-order derivative is calculated, $I'(g_i)$ along $0°$, $45°$, $90°$ and $135°$ directions. Figure 7.1 explains the detailed calculation of DBC along $0°$ direction.

#### 7.8.2.3 Directional Local Motif XoR Patterns

The image is divided into an overlapped 3×3 grid. From each grid, four 1×3 sub-grids are retrieved in $0°$, $45°$, $90°$ and $135°$ directions. All four sub-grids are constructed based on the center pixel of 3×3 grid. The complete details are given in S. K. Vipparthi and Nagar (2014b).

### 7.8.3 PROPOSED FEATURE DESCRIPTORS

In multi-directional local motif XoR patterns (MDLMXoRP) first, the directional information along $0°$ are coded. Further, expert smart grid XoR operation is coded on the resulting directional image. The complete operation is given in Figure 7.48. This entire texture calculation is done in V pattern of the HSV color space. Two different quantization levels of S and V planes are also tested on directional image. The complete proposed system framework is also shown in Figure 7.46.

#### 7.8.3.1 Analysis

Here our main aim is to determine a new transformed motif number based on the pixel behavior. The detailed explanation for $0°$ direction motif calculation is given as follows:

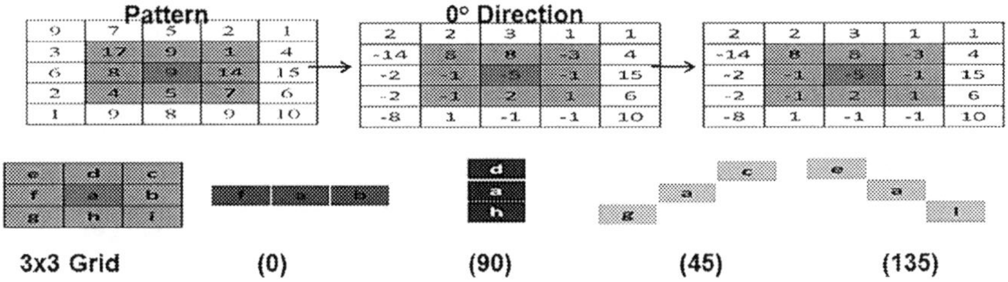

| Pattern | | | | | | 0° Direction | | | | | | | | | | |
|---|---|---|---|---|---|---|---|---|---|---|---|---|---|---|---|---|
| 9 | 7 | 5 | 2 | 1 | | 2 | 2 | 3 | 1 | 1 | | 2 | 2 | 3 | 1 | 1 |
| 3 | 17 | 9 | 1 | 4 | | -14 | 8 | 8 | -3 | 4 | | -14 | 8 | 8 | -3 | 4 |
| 6 | 8 | 9 | 14 | 15 | | -2 | -1 | -5 | -1 | 15 | | -2 | -1 | -5 | -1 | 15 |
| 2 | 4 | 5 | 7 | 6 | | -2 | -1 | 2 | 1 | 6 | | -2 | -1 | 2 | 1 | 6 |
| 1 | 9 | 8 | 9 | 10 | | -8 | 1 | -1 | -1 | 10 | | -8 | 1 | -1 | -1 | 10 |

3x3 Grid     (0)     (90)     (45)     (135)

**FIGURE 7.46** Smart grid retrieval from 0° directional image.

Let a, b and c be the gray values of 1×3 sub-grid in 0° direction (see in Figure 7.2).

1. **For motif "1"**: when a=1, b= 2 and C=3. Here, "b" is the center pixel value in 1×3 pattern. If it satisfies the condition of "b>a" and "b<C" then the 1×3 pattern is replaced by a new motif value '1'. Similarly, motif "2" is coded when "A>b" and "b>c".

2. **For motif "3"**: when "A"=4, "b"=3 and "C" = 5. Center pixel value is smaller than the neighbors and "A<C". Then, the new motif value with "3" is coded. Similarly, motif "4" is coded when "A>C".

3. **For motif "5"**: as if "a"=3, "b"=5 and "c" =4. Center pixel value is larger than the neighbors and "a<c". Then, new motif value with "5" is coded. Similarly, motif "6" is coded when "a>c".

4. **For motif "7"**: if all three pixels bear same gray-scale values, then, new motif value with "7" is coded.

In this chapter, following notations are used for better understanding. Smaller and larger value is defined by using "<" and ">" symbols. Whereas "<<" and ">>" symbols are referred to smaller of the smallest and larger of the largest values, respectively, among "a" and "c" pixels. If "a" value is larger of the largest value then superscript of "$\bar{A}$" and ">>" symbol are used. Similarly for smaller of the smallest value "<<" and subscript of "$\underline{c}$" is used and vice versa in Figure 7.47.

$$T_{\text{Im}}^{\theta_1}(a,b,c) = \begin{cases} 1, & \left((a \geq d)\&(a \leq b)\right) \\ 2, & \left((a \leq d)\&(a \geq b)\right) \\ 3, & \left((a \leq d)\&(a \leq b)\&(d \leq b)\right) \\ 4, & \left((a \leq d)\&(a \leq b)\&(b \leq d)\right) \\ 5, & \left((a \geq d)\&(a \geq b)\&(d \leq b)\right) \\ 6, & \left((a \geq d)\&(a \geq b)\&(b \leq d)\right) \\ 7, & \left((a = d)\&(a = b)\&(b = d)\right) \end{cases}_{\theta_1 = 0°,90°,180°,270°} \qquad (7.89)$$

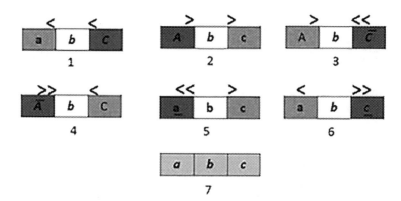

**FIGURE 7.47**   Illustration of motif computation in smart grid.

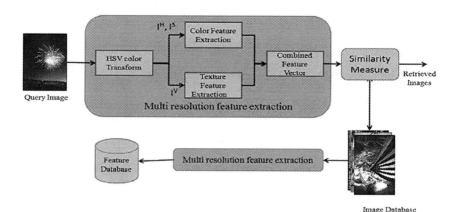

**FIGURE 7.48**   The proposed image retrieval system framework.

Finally, the whole image is converted into four directional transformed motif images Figure 7.48.

For a given 3×3 grid, MDLMXoRP value is computed by comparing its center motif value with its surrounding motif values using Equation (7.89).

$$I'(g_i) = I(g_c) - I(g_i); \quad i = 1, 2, 3, 4 \tag{7.90}$$

The directional edges are obtained by

$$\hat{I}_\alpha^{DIM}(g_c) = f_4(I'(g_j)); \quad j = (1 + \alpha/45) \ \forall \alpha = 0°, 45°, 90° \text{ and } 135° \tag{7.91}$$

$$D_{IM}(I(g_c))\big|_\alpha = \left\{ \hat{I}_\alpha^{DIM}(g_c); \hat{I}_\alpha^{DIM}(g_1); \hat{I}_\alpha^{DIM}(g_2); \ldots \ldots \hat{I}_\alpha^{DIM}(g_8) \right\} \tag{7.92}$$

$$D_{k,j,lm}\left(p_i,p_c\right)=\begin{cases}1 & p_i\neq p_c \\ 0 & p_i=p_c\end{cases}\forall k=1,2,3,4\,\&\,j=1 \qquad (7.93)$$

where, N is number of neighbors, R is radius of the neighbor, $p_c$ denotes motif value of a center pixel in a given motif image, $p_i$ is the motif value of neighborhood.

After computing the local motif XOR pattern (LMXoRP) for each pixel (i, j) in four directions, the whole image is represented by building a histogram as follows:

$$H_{k,j,\text{MDLMXoRPs}}\left(l\right)=\sum_{i=1}^{N_1}\sum_{j=1}^{N_2}f\left(\text{MDLMXoRP}^v\left(i,j\right),l\right);l\in\left[0,\left(2^p-1\right)\right];$$
$$\forall k=1,2,3,4\,\&\,j=1 \qquad (7.94)$$

$$f\left(x,y\right)=\begin{cases}1 & x=y \\ 0 & else\end{cases} \qquad (7.95)$$

where size of the input image is $N_1 \times N_2$.

The overall histogram for all five MDLMXoRP is concatenated to form the proposed CMDLMXoRPs descriptor of the input image.

$$H_{\text{MDLMXoRP}}=\left[H^v_{\text{1MDLMXoRP}};H^H;H^S\right] \qquad (7.96)$$

The Proposed System Frame Work

The flowchart of the proposed image retrieval system and its algorithm is given below.

**ALGORITHM**

*Input: Color Image; Output: Retrieval Results*

1. Load the color RGB image and transform into an HSV color image.
2. Collect the IV component from the HSV color space.
3. Extract 0° information from IV color component.
4. Divide the image into 3×3 grids.
5. Collect four 1×3 sub-grids from each 3×3 grid.
6. Calculate the four motif images in 0°, 45°, 90° and 135° directions.
7. Apply XOR operation on four motif images to form MDLMXoRP.
8. Construct a feature vector by concatenating the histograms of H, S components with V component.
9. Compare the query feature with features in the database.
10. Retrieve the images based on the best matches.

Similarity Distance Measures

In this chapter, the performance of the proposed method is analyzed with $d_1$ distance similarity measures as shown in Equation (7.97).

$$d_1 \, Distance: D_s\left(Q_m, T_m\right) = \sum_{i=1}^{Lg} \left| \frac{f_{T_m,i} - f_{Q_m,i}}{1 + f_{T_m,i} + f_{Q_m,i}} \right| \tag{7.97}$$

where $Q_m$ represents the query image, $Lg$ is the feature vector length, $T_m$ is image in database; $f_{Tm,i}$ is the $i^{th}$ feature of image $T_m$ in the database, $f_{Qm,i}$ is the $i^{th}$ feature of query image $Q_m$.

### 7.8.4   EXPERIMENTAL RESULTS AND DISCUSSIONS

The performance of the proposed method is tested by conducting two experiments on benchmark databases. By domain professionals this database is pre-classified into different categories of size 100. Because of its size and heterogeneous content Corel-database meets all the requirements to evaluate an image retrieval system. In this chapter, we use the Corel-5000 and Corel-10000 databases which consist of 50 and 100 different categories, respectively. Each category has $N_G$ (=100) images with resolution of either $126 \times 187$ or $187 \times 126$.

The retrieval performance of the proposed method measured in terms of precision (P), ARP, recall (R) and ARR are shown below.

For the query image $I_q$ the precision (P) and Recall (R) are defined as follows:

$$Precision: P\left(Iq\right) = \frac{Number \, of \, Relevant \, Images \, Retrieved}{Total \, Number \, of \, Images \, Retrieved} \tag{7.98}$$

$$ARP = \frac{1}{DB} \sum_{i=1}^{DB} P\left(Ii\right) | n \leq 10 \tag{7.99}$$

$$Recall: R\left(Iq\right) = \frac{Number \, of \, Relevant \, Images \, Retrieved}{Total \, Number \, of \, Relevant \, Images \, in \, the \, database} \tag{7.100}$$

$$ARR = \frac{1}{DB} \sum_{i=1}^{DB} R\left(Ii\right) | n \geq 10 \tag{7.101}$$

where DB is the total number of images in the database.

### 7.8.4.1   Experiment #1

In this experiment, the performance of the proposed method (MDLMXoRP) is tested on Corel-5000 database in terms of ARP and ARR. The retrieval performance of the proposed method and other existing methods are shown in Figure 7.49(a) and 7.49(b)

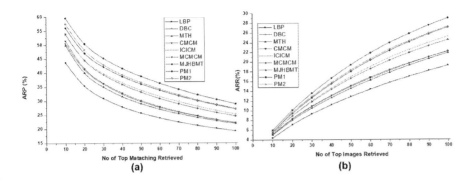

**FIGURE 7.49**   The retrieval performance of the proposed method in terms of ARP and ARR on Corel-5K database.

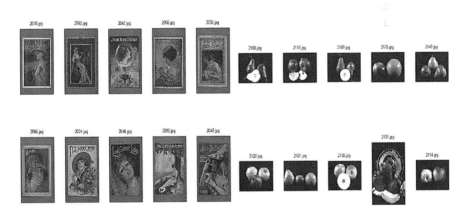

**FIGURE 7.50**   Two query results of the proposed method on Corel-5K database (top-left image is the query image).

in terms of ARP and ARR, respectively, on Corel 5K database. From Figure 7.49 it is very clear that the proposed method with less quantization level has shown better performance than other state-of-the-art techniques. The query results of the proposed method on Corel-5000 database is (top-left image is query image) shown in Figure 7.50.

### 7.8.4.2   Experiment #2

In this experiment, Corel-10000 database is used to evaluate the performance of the proposed method. The retrieval performance of the proposed method and other existing methods are shown in Figure 7.51(a) and 7.52(b) in terms of ARP and ARR, respectively, on Corel 10K database. From Figure 7.51 it is very clear that the proposed method with less quantization level has shown better performance than other

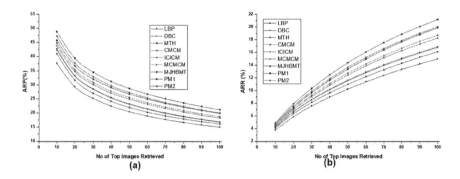

**FIGURE 7.51** The retrieval performance of the proposed method in terms of ARP and ARR on Corel-10K database.

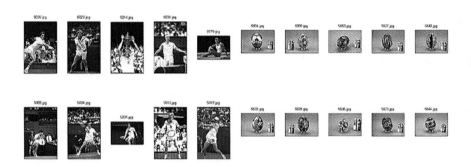

**FIGURE 7.52** Two query results of the proposed method on Corel-10K database (top-left image is the query image).

state-of-the-art techniques. The query results of the proposed method on Corel-10000 database is (top-left image is query image) shown in Figure 7.52.

### 7.8.5 Conclusion

In this chapter, a simple and more efficient feature extraction approach called Color-Based Multi-Directional Local Motif XoR Patterns (MDLMXoRP) is proposed for image retrieval. The MDLMXoRP collects the relationship between the directional smart grids using joint correlation histograms. Further, the effectiveness of the proposed method is improved by integrating it the HSV color histogram features. For color and texture integration, we use the H and S components for color histogram calculation and V component for MDLMXoRP features extraction. The performance of the proposed method is tested on Corel-5K and Corel-10K databases. The retrieval results after investigation show significant improvements in terms of their evaluation measures as compared to the state-of-the-art techniques for image retrieval on respective databases.

## 7.9 QUANTIZED LOCAL TRIO PATTERNS FOR MULTIMEDIA IMAGE RETRIEVAL SYSTEM

### 7.9.1 INTRODUCTION

CBIR describes the image on the basis of explored parameters. It extracts the features from visual part of the image such as texture, shape and color. Contributions made for the improvement in the efficacy of retrieval systems are given in M. L. Kherfi et al. (2004), R. Dutta et al. (2008). Texture is one of the primary features used in CBIR. Researchers proposed many retrieval methods by using the texture as the primary element V. N. Gudivada and Raghavan (1995), A. W. M. Smeulders et al. (2000), H. Muller et al. (2001), R. Veltkamp and Tanase (2002), M. L. Kherfi et al. (2004). L. B. P. Ojala et al. (2002) became the foundation for a few retrieval algorithms. Takala et al. (2005) introduced block-based L. B. P. Yao and Chen (2003) proposed LEPSEG, LEPINV for texture description. Concepts of local trio patterns S. K. Vipparthi et al. (2015) and quantized patterns K. Rao et al. (2015) made us to introduce a feature vector named quantized local trio patterns binary pattern (QLTP) for image retrieval. Primary contributions to the devised method are as follows: 1) New feature in the form of a pattern (QLTP) integrates the quantization and local trio patterns for image retrieval. 2) The proposed method collects the texture information by extracting the relationship between pixels. 3) Retrieval ability of our proposed descriptor is evaluated using benchmark Corel-10K database. The retrieval results show significant improvement when compared to some of the recent methods of image retrieval. Contents in this chapter are organized as follows: Section 7.9.1 provides an introduction of CBIR. Section 7.9.2 present local extreme sign trio pattern patterns. The proposed descriptor is discussed in Section 7.9.3. Results of the experiments using different similarity measures are given in Section 7.9.4. Section 7.9.5 covers the conclusions of the work.

### 7.9.2 LOCAL EXTREME SIGN TRIO PATTERN

S. K. Vipparthi and Nagar (2014a) proposed a pattern named LESTP for an image retrieval system. LESTP extracts the extreme sign information from an image using trio values. An extreme edge is extracted by using the sign of a difference between the central pixel and all its neighbors located within the radius of one position. Detailed presentation on LESTP is available in S. K. Vipparthi and Nagar (2014a).

### 7.9.3 PROPOSED METHOD

Motivated by local trio patterns, we introduce a new feature descriptor by applying the quantization to the image as specified in Equations (7.102)–(7.105). Trio patterns are obtained according to Equation (7.106).

$$DQE(I(p_c))\big|_{0°} = \left\{ g(I(p_1), I(p_4), I(p_C)); g(I(p_2), I(p_5), I(p_C)); g(I(p_3), I(p_6), I(p_C)) \right\}$$

$$(7.102)$$

$$DQE(I(p_c))\big|_{45°} =$$
$$\{g(I(p_{13}),I(p_{16}),I(p_c)); g(I(p_{14}),I(p_{17}),I(p_c)); g(I(p_{15}),I(p_{18}),I(p_c))\} \quad (7.103)$$

$$DQE(I(p_c))\big|_{90°}$$
$$= \{g(I(p_7),I(p_{10}),I(p_C)); g(I(p_8),I(p_{11}),I(p_C)); g(I(p_9),I(p_{12}),I(p_C))\} \quad (7.104)$$

$$DQE(I(p_c))\big|_{135°}$$
$$= \{g(I(p_{19}),I(p_{22}),I(p_C)); g(I(p_{20}),I(p_{23}),I(p_C)); g(I(p_{21}),I(p_{24}),I(p_C))\} \quad (7.105)$$

where

$$g(t_1,t_2,p) = \begin{cases} 1 & if\left(p >= t_1\right) \\ -1 & if\left(p1 <= t_2\right) \\ 0 & if\left(t1 > p < t_2\right) \end{cases} \quad (7.106)$$

The process of quantization is depicted in Figure 7.1 and Figure 7.2 illustrates the calculation of QLTP for the pixel at the center. Intensity variation between the central pixel and neighbor pixels on both the sides are "18, 12, 7, −7, −3, −2" which are arranged in decreasing order based on the magnitudes. The first value indicates the extreme edge of pixel of high value. These edge values are converted into trio, (−1, 0, 1), values using two threshold limits $\tau_1$ and $\tau_2$. At present, we consider thresholds as $\tau_1 = -3$ and $\tau_2 = 3$. According to (7), the conversion QLTP is represented as 1, 1, 1, −1, −1, 0. Similarly, the procedure is followed in case of all neighboring pixels. Finally, the QLTP is converted into one upper and one lower pattern. This pattern shows a considerable improvement than other state-of-the-art techniques. QLTP extracts more information pertaining to an edge as compared to the popular LBP.

In our proposed method, the given image is quantized to get HVDA structure as shown in Figure 7.53. The HVDA reflects the stretch of pixels in horizontal, vertical, diagonal and anti-diagonal directions.

Upon quantizing the image in different directions, the trio patterns are extracted in all the four directions of the quantized structure. For instance, extraction of trio patterns for the vertical strip of pixels is shown in Figure 7.54.

| 22 | 46 | 41 | 38 | 42 | 43 | 12 |
|----|----|----|----|----|----|----|
| 44 | 32 | 45 | 13 | 46 | 32 | 47 |
| 47 | 33 | 16 | 18 | 25 | 30 | 46 |
| 21 | 18 | 17 | 20 | 11 | 24 | 28 |
| 34 | 39 | 11 | 17 | 37 | 54 | 59 |
| 45 | 48 | 37 | 32 | 42 | 41 | 38 |
| 33 | 60 | 61 | 27 | 62 | 63 | 39 |

| 22 |    |    | 38 |    |    | 12 |
|----|----|----|----|----|----|----|
|    | 32 |    | 13 |    | 32 |    |
|    |    | 16 | 18 | 25 |    |    |
| 21 | 18 | 17 | 20 | 11 | 24 | 28 |
|    |    | 11 | 17 | 37 |    |    |
|    | 48 |    | 32 |    | 41 |    |
| 33 |    |    | 27 |    |    | 39 |

**FIGURE 7.53**    A 7 × 7 image and the quantized structure.

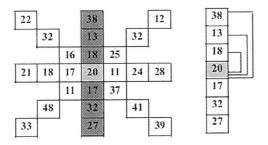

| 18 | 12 | 7 | -7 | -3 | -2 |
|----|----|----|----|----|----|
| 1 | 1 | 1 | -1 | -1 | 0 |
| 1 | 1 | 1 | 0 | 0 | 0 |
| 0 | 0 | 0 | 1 | 1 | 0 |

**FIGURE 7.54** Calculation of the proposed pattern from the quantized structure: Threshold $(t_1) = 3$, Threshold $(t_2) = -3$.

## ALGORITHM

Input: Image Output: Retrieval result

1. Convert the color image to gray-scale image.
2. Extract the quantized patterns for the gray-scale image.
3. Collect the trio patterns by setting any two threshold values.
4. Derive the ULESTP, LLESTP from the strips of pixels.
5. Construct a histogram to create the final feature vector.
6. Compare query image with database images.
7. Retrieve the images according to the best matches.

### 7.9.4 EXPERIMENTAL RESULTS AND DISCUSSIONS

Efficacy of the proposed method is determined by means of the experiments on standard databases. During the experimentation, each database image becomes a query image. Accordingly, the proposed work identifies $n$ images $Y=(y_1, y_2, ...,y_n)$ according to the shortest distance. Performance is evaluated in terms of average precision (ARP) and average recall (ARR) as indicated below: Evaluation metrics for a query image $Im_q$ are determined basing on the equation given below.

$$Pre: P\left(Im_q\right) = \frac{No.\, of\ related\ images\ retrieved}{Total\ images\ retrieved} \tag{7.107}$$

**TABLE 7.12**

**Precision and Recall Values of Various Methods (Corel-10k)**

| Image Repository | Evaluation Metric | CS_LBP | LEP SEG | LEP INV | BLK_LBP | DLEP | LECTP | QLTP |
|---|---|---|---|---|---|---|---|---|
| Corel-10k | %Prec | 26.4 | 34.0 | 28.9 | 38.1 | 39.9 | 28.3 | 29.2 |
| | %Recall | 10.1 | 13.8 | 11.2 | 15.3 | 157 | 16.6 | 18.2 |

$$Avg.\ Retrieval\ Prec: ARP = \frac{1}{|DBS|} \sum_{k=1}^{|DBS|} \Pr\left(Im_k\right)$$  (7.108)

$$Recall: Re\left(Im_q\right) = \frac{No.\ of\ related\ images\ retrieved}{Total\ relevant\ images\ in\ database}$$  (7.109)

$$Avg.\ Retreival\ Rate: ARR = \frac{1}{|DBS|} \sum_{k=1}^{|DBS|} \Re\left(I_k\right)$$  (7.110)

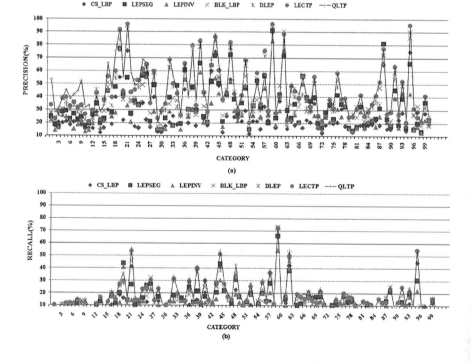

**FIGURE 7.55**   Illustration of various methods in the form of (a) ARP (b) ARR.

### 7.9.4.1 Corel-10k

Corel-10k repository is an agglomeration of 10000 images classified into different categories. Performance of QLTP is determined by means of Precision, Recall, ARR and ARP. Precision and recall of retrieval are given in Table 7.12.

Comparisons of the results are shown in Figure 7.55. The improvement in performance is primarily due to the discerning capability of the proposed method against any other method mentioned.

### 7.9.5 Conclusion

A new method for image retrieval is introduced in this chapter. It explores the quantized trio patterns for image retrieval. Corel-10k repository is used to test the performance of QLTP method. Results after investigation exhibit a substantial improvement in the values of precision as well as recall.

By extracting the facial features of a given image, the proposed QLTP framework can be extended in the field of face recognition.

# 8 Conclusion and Future Scope

## 8.1 SUMMARY

Large volumes of data in the form of text, images, videos, files, etc., are available everywhere. Due to the evolutionary changes in the World Wide Web, image retrieval has attained wide popularity in recent times. Traditional retrieval methods totally depend on human annotation of images which has some limitations. Content-based image retrieval is an automated method that involves extraction of inherent content of the image. In this method, feature vectors generated using a single pixel provide less information as compared to a group of pixels. Correlation among the pixels of a neighborhood represents more useful information about the structure of texture. Local binary patterns are proved to be effective in deriving texture information in an effective manner for any practical system. However, the size of feature vector becomes large when the neighborhood increases.

Based on the concept of local binary patterns, various types of feature vectors are introduced in this book to enhance performance of the retrieval system in the present context. Local extremas are calculated in increased neighborhood of a pixel in the image. Color and texture are integrated with directional local extrema patterns to enhance performance.

Quantization of extrema is implemented to increase the accuracy of the system for larger databases. As a variant of quantized extrema patterns, Color information is collected from two oppugnant planes. A mesh is created out of the extrema pattern to propose mesh quantized extrema patterns. Apart from this, a few more patterns are proposed for image retrieval. Different types of image databases such as Corel-1k, Corel-5k, Corel-10k, ImageNet-25k and MIT VisTex are used to test the effectiveness of the proposed methods. Euclidean distance, D1 distance, Manhattan distance and Canberra distance are used for comparison purposes.

## 8.2 SALIENT FEATURES

### 8.2.1 IMPROVED DIRECTIONAL LOCAL EXTREMA PATTERNS

- Directional local extrema patterns are integrated with color and Gabor features to create Color Directional Local Extrema Patterns (CDLEP) and Gabor directional local extrema patterns, respectively.
- These models are tested on Corel-1k repository which has shown satisfactory improvement in retrieval results.
- Average precision and average recall values for Corel-1k database are 75.5% and 50.4%, respectively, in case of CDLEP.

DOI: 10.1201/9781003123514-8

- Average precision and average recall values for Corel-1k database are 75.9% and 50.0%, respectively, in case of Gabor directional local extrema patterns.

### 8.2.2 LOCAL QUANTIZED EXTREMA PATTERNS

- Quantization is applied on extrema patterns to create local quantized extrema patterns.
- Local quantized extrema patterns extract textural structure according to the difference in intensity of the center pixel and the neighbors in four directions.
- Color RGB histogram is combined with local quantized extrema to obtain the final form of descriptor.
- Corel-1k, Corel-5k and MIT VisTex image repositories are used to test the effectiveness of the method.
- Euclidean distance, d1 distance, Manhattan distance and Canberra distance are used for comparison purpose.
- Average precision and Average recall values for Corel-1k database are 76.7% and 54.3%, respectively. For Corel-5k, average precision and average recall are 62.2% and 29.6%, respectively. In case of MIT VisTex, the values of average precision and average recall are 99.4% and 58.3%, respectively.

### 8.2.3 LOCAL COLOR OPPUGNANT QUANTIZED EXTREMA PATTERNS

- Local color oppugnant quantized extrema patterns approach extracts textural structure from two different color planes according to the difference in intensity of the center pixel and the neighbors in four directions.
- Information from RGB and HSV is used to generate feature vector.
- Corel-1k, Corel-5k, Corel-10k and ImageNet-25k image repositories are used to test performance.
- Euclidean distance, d1 distance, Manhattan distance and Canberra distance are used for comparison purpose.
- Average precision and average recall values for Corel-1k database are 78.4% and 57.8%, respectively. For Corel-5k, average precision and average recall are 59.2% and 27.4%, respectively. In case of Corel-10k, the values of average precision and average recall are 49.4% and 20.9%, respectively. For ImageNet-25k, average precision is 36.4%.

### 8.2.4 LOCAL MESH QUANTIZED EXTREMA PATTERNS

- A mesh is formed using the picture elements at alternate positions of Horizontal vertical diagonal anti-diagonal structure.
- Color information in the form of RGB is combined to generate the feature descriptor.
- Corel-1k, MIT VisTex repositories are used to test performance of this approach. Average precision for Corel-1k database is 88.5% and for MIT Vis Tex database, average recall value is 59.2%.

- From these results, it is established that proposed methods exhibit satisfactory improvement in average precision and average recall values. Dimensionality of feature vector in these models is maintained at 4096 while achieving improved accuracy.

## 8.3   FUTURE SCOPE

- Improvement of accuracy of the system while minimizing dimensionality of feature vector is an open issue.
- Present work is limited to natural and texture databases with varying information. These models can be applied to data containing boundary information such as facial images where the amount of information is less than natural images.

# References

Abrishami-Moghadam, H., T. T. Khajoie, A. H. Rouhi, and M. Saadatmand, "Wavelet correlogram: A new approach for image indexing and retrieval," *Journal of Pattern Recognition*, Vol. 38, 2005, pp. 2506–2518.

Abrishami-Moghadam, H., A. H. Rouhi, and T. Taghizadeh Khajoie, *"A new algorithm for image indexing and retrieval using wavelet correlogram,"* presented at the *IEEE International Conference on Image Processing (ICIP)*, Spain, 2003.

Ahmad Fauzi, M. F. and P. H. Lewis, "A multiscale approach to texture-based image retrieval," *Pattern Analysis and Application*, Vol. 11, 2008, pp. 141–157.

Ahmadian, A. and A. Mostafa, *"An efficient texture classification algorithm using Gabor wavelet,"* in *Proceedings of the 25th Annual International Conference of the IEEE Engineering in Medicine and Biology*, Cancun, Mexico, 2003, pp. 930–933.

Ahmadian, A., A. Mostafa, M. Abolhassani, and Y. Salimpour, "A texture classification method for diffused liver diseases using Gabor wavelets. Annual International Conference of the IEEE Engineering in Medicine and Biology Society," *IEEE Engineering in Medicine and Biology Society Conference*, Vol. 2(c), 2005, pp. 1567–1570. doi:10.1109/IEMBS. 2005.1616734

Ahmed, S., M. Weber, M. Liwicki, C. Langenhan, A. Dengel, and F. Petzold, "Automatic analysis and sketch-based retrieval of architectural floor plans," *Pattern Recognition Letters*, Vol. 35, 2014, pp. 91–100.

Ahonen, T., A. Hadid, and M. Pietikainen, "Face description with local binary patterns: Applications to face recognition," *IEEE Transactions on Pattern Analysis and Machine Intelligence*, Vol. 28(12), 2006, pp. 2037–2041.

Almeida, L. B., "The fractional Fourier transform and time-frequency representations," *IEEE Transactions on Signal Processing*, Vol. 42, 1994, pp. 3084–3091.

Andrews-Hanna, J. R., R. Saxe, and T. Yarkoni, "Contributions of episodic retrieval and mentalizing to autobiographical thought: Evidence from functional neuroimaging, resting-state connectivity, and fMRI meta-analyses," *NeuroImage*, Vol. 91, 2014, pp. 324–335.

Anuar, F. M., R. Setchi, and Y-k. Lai, "Trademark image retrieval using an integrated shape descriptor," *Expert Systems with Applications*, Vol. 40, 2013, pp. 105–121.

Arivazhagan, S. and L. Ganesan, "Texture classification using wavelet transform (1513–1521)," *Pattern Recognition Letters*, Vol. 24(9–10), 2003.

Arivazhagan, S. and L. Ganesan, "Texture classification using Gabor wavelets based rotation invariant features," *Pattern Recognition Letters*, Vol. 27(6), 2006, pp. 1976–1982.

Banerjee, P., A. K. Bhunia, A. Bhattacharyya, P. P. Roy, and S. Murala, "Local neighborhood Intensity pattern-a new texture feature descriptor for image retrieval," *Expert Systems with Applications*, Vol. 113, 2018, pp. 100–115.

Barnachon, M., S. Bouakaz, B. Boufama, and E. Guillou, "A real-time system for motion retrieval and interpretation," *Pattern Recognition Letters*, Vol. 34, 2013, pp. 1789–1798.

Chang, M. H., J. Y. Pyu, M. B. Ahmad, J. H. Chun, and J. A. Park, "Modified color co-occurrence matrix for image retrieval," *Advances in Natural Computation*, 2005, pp. 43–50.

Chen, B.-C., Y.-Y. Chen, Y.-H. Kuo, and W. H. Hsu, "Scalable face image retrieval using attribute-enhanced sparse codewords," *IEEE Transactions on Multimedia*, Vol. 15(5), 2013, pp. 1163–1173.

Chen, Q., J. C. Luo, and C. Zhou, "Multispectral satellite imagery segmentation using a simplified JSEG approach," *Applications of Digital Image Processing XXVII*, Vol. 5558, 2004, pp. 853–861.

Chikano, M., K. Kise, M. Iwamura, S. Uchida, and S. Omachi, "Recovery and localization of handwritings by a camera-pen based on tracking and document image retrieval," *Pattern Recognition Letters*, Vol. 35, 2014, pp. 214–224.

Cinque, L., S. Levialdi, and A. Pellicano, "Color-based image retrieval using spatial-chromatichistograms," in *IEEE International Conference on Multimedia Computing and Systems*, 1999, pp. 969–973.

Ciresan, D., U. Meier, and J. Schmidhuber, "*Multi-column deep neural networks for image classification*," in *Computer Vision and Pattern Recognition (CVPR), 2012 IEEE Conference on IEEE*, 2012, pp. 3642–3649.

Conci, A. and E. M. M. M. Castro, "Image mining by content," *Expert Systems with Applications*, Vol. 23(4), 2002, pp. 377–383.

Corel 1000 and Corel 10000 image database. [Online]. Available: http://wang.ist.psu.edu/docs/related.shtml.

Corel Database: http://www.ci.gxnu.edu.cn/cbir/Dataset.aspx

Do, M. N. and M. Vetterli "Wavelet-based texture retrieval using generalized Gaussian density and Kullback-leibler distance," *IEEE Transactions on Image Process*, Vol. 11, 2002, pp. 146–158.

Dohnal, V., C. Gennaro, and P. Zezula, "D-Index: Distance searching index for metric data sets," *Multimedia Tools and Applications*, Vol. 21, 2003, pp. 9–33.

Donoho, D. L. and M. R. Duncan, "Digital curvelet transform: Strategy implementation and experiments," in *Aerosense 2000, Wavelet Applications VII*, Bellingham, Washington: SPIE, 2000, pp. 12–29.

Dubey, S. R., S. K. Singh, and R. K. Singh, "Local diagonal extrema pattern: A new and efficient feature descriptor for CT image retrieval," *IEEE Signal Processing Letters*, Vol. 22(9), 2015, pp. 1215–1219.

Dutta, R., D. Joshi, J. Li, and J. Z. Wang, "Image retrieval: Ideas, Influences and trends of the new age," *ACM Computing Surveys*, Vol. 40(2), 2008, pp. 1–60.

Espinoza-Molina, D. and M. Datcu, "Earth-observation image retrieval based on content, semantics, and metadata," *IEEE Transactions on Geoscience and Remote Sensing*, Vol. 51(11), 2013, pp. 5145–5159.

Evans, L. H., A. N. Williams, and E. L. Wilding, "Electrophysiological evidence for retrieval mode immediately after a task switch," *NeuroImage*, Vol. 108, 2015, pp. 435–440.

Everingham, M., S. M. A. Eslami, L. Van Gool, C. K. I. Williams, J. Winn, and A. Zisserman, "The PASCAL visual object classes challenge: A retrospective," *International Journal of Computer Vision*, Vol. 111(1), 2015, pp. 98–136.

Fadili, M. J. and J. L. Starck, *Curvelets and Ridgelets, Encyclopedia of Complexity and System Science*. New York: Springer, 2007.

Fakheri, M., T. Sedghi, M. G. Shayesteh, and M. C. Amirani, "Framework for image retrieval using machine learning and statistical similarity matching techniques," *IET Image Processing*, Vol. 7(1), 2013, pp. 1–11.

Frosini, P. and C. Landi, "Persistent Betti numbers for a noise tolerant shape-based approach to image retrieval," *Pattern Recognition Letters*, Vol. 34, 2013, pp. 863–872.

Galshetwar, G. M. et al., "Edgy salient local binary patterns in inter-plane relationship for image retrieval in Diabetic Retinopathy," *Procedia Computer Science*, Vol. 115, pp. 440–447. doi:10.1016/j.procs.2017.09.103

Gonde, A. B., R. P. Maheshwari, and R. Balasubramanian, "Complex Wavelet Transform with Vocabulary Tree for content based image retrieval", Proc. Seventh Indian Conference on Computer Vision, Graphics and Image Processing (ICVGIP'10), Chennai, pp. 359–366, Dec 2010.

Gonde, A. B., S. Murala, S. K. Vipparthi, R. Maheshwari, and R. Balasubramanian, "3D local transform patterns: A new feature descriptor for image retrieval," In B. Raman, S. Kumar, P. Roy, and D. Sen, eds. *Proceedings of International Conference on Computer Vision and Image Processing. Advances in Intelligent Systems and Computing*, Vol. 460, 2017, Singapore: Springer.

Gonzalez-Diaz, I., C. E. Baz-Hormigos, and F. Diaz-de-Maria, "A generative model for concurrent image retrieval and ROI segmentation," *IEEE Transactions on Multimedia*, Vol. 16(1), 2014, pp. 169–183.

Gordo, A., F. Perronnin, and E. Valveny, "Large-scale document image retrieval and classification with runlength histograms and binary embeddings," *Pattern Recognition*, Vol. 46, 2013, pp. 1898–1905.

Gudivada, V. N. and V. V. Raghavan, "Content-based image retrieval," *Computer*, Vol. 28, 1995, pp. 18–22.

Guo, X. C. and D. Hatzinakos, "Content based image hashing via wavelet and radon transform," *Lecture notes in Computer Science*, Vol. 4810, 2007, pp. 755–764.

Guo, J.-M. and H. Prasetyo, "Content-based image retrieval using features extracted from Halftoning-based block truncation coding," *IEEE Transactions on Image Processing*, Vol. 24(3), 2015, pp. 1010–1024.

Guo, Z., L. Zhang, and D. Zhang, "A completed modeling of local binary pattern operator for texture classification," *IEEE Transactions on Image Processing: A Publication of the IEEE Signal Processing Society*, Vol. 19(6), 2010, pp. 1657–1663. doi:10.1109/TIP. 2010.2044957

Hadid, A., G. Zhao, T. Ahonen, and M. Pietikainen, "Face analysis using local binary patterns," in *Handbook of Texture Analysis*, London: Imperial College Press, 2008, pp. 347–373.

Han, J. and K. K. Ma, "Rotation-invariant and scale-invariant Gabor features for texture image retrieval," *Image and Vision Computing*, Vol. 25, 2007, pp. 1474–1481.

Han, J. and S. J. McKenna, "Query-dependent metric learning for adaptive, content-based image browsing and retrieval," *IET Image Processing*, Vol. 8(10), 2014, pp. 610–618.

Haralick, R. M., "Statistical and structural approaches to texture," *Proceedings IEEE*, Vol. 67(5), 1979, pp. 786–804.

Haralick, R. M., K. Shanmugam, and I. Dinstein, "Textural features for image classification," *IEEE Transactions on Systems, Man and Cybernatics*, Vol. 3(6), 1973a, pp. 610–621.

Haralick, R. M., K. Shanmugam, and I. Dinstein, "Texture features for image classification," *IEEE Transactions on System, Man and Cybernetics*, Vol. smc-8, 1973b, pp. 610–621.

He, D-C. and L. Wang, "Texture unit, texture spectrum and Texture," *IEEE Transactions on Geoscience and Remote Sensing*, Vol. 28, 1990.

Heikkila, M. and M. Pietikainen, "A texture-based method for modelling the background and detecting moving objects," *IEEE Transactions on Pattern Analysis and Machine Intelligence*, 2006, pp. 657–662.

Heikkila, M., M. Pietikainen, and C. Schmid, "Description of interest regions with local binary patterns," *Pattern Recognition*, Vol. 42(3), 2009, pp. 425–436.

Huang, J., S. R. Kumar and M. Mitra, "*Combining supervised learning with color correlograms for content-based image retrieval*," in *Proceedings of the 5th ACM Multimedia Conference*, 1997, pp. 325–334.

Huang, X., S. Z. Li, and Y. Wang, "Shape localization based on statistical method using extended local binary patterns," *Proceedings of the International Conference on Image and Graphics*, 2004, pp. 184–187.

Huijsmans, D. P. and N. Sebe, "How to complete performance graphs in content-based image retrieval: Add generality and normalize scope," *IEEE Transactions on Pattern Analysis and Machine Intelligence*, Vol. 27(2), 2005, pp. 245–251.

Hussain, S. and B. Triggs, "Visual recognition using local quantized patterns," ECCV 2012, Part II, LNCS 7573, Italy, 2012, pp. 716–729.

Idris, F. and S. Panchanathan, "Image and video indexing using vector quantization," *Machine Vision and Applications*, Vol. 10, 1997, pp. 43–50.

ImageNet database available. online at imagenet.org

Jabid, T., H. Kabir, and O. Chae, "*Local Directional Pattern for face recognition*" IEEE conference 2010. http://wang.ist.psu.edu/docs/related/

Jeena Jacob, I., K. G. Srinivasagan, and K. Jayapriya, "Local Oppugnant color texture pattern for image retrieval system," *Pattern Recognition Letters*, Vol. 42, 2014, pp. 72–78.

Jégou, H., F. Perronnin, M. Douze, J. Sanchez, P. Perez, and C. Schmid, "Aggregating local image descriptors into compact codes. Pattern Analysis and Machine Intelligence," *IEEE Transactions on*, Vol. 34(9), 2012, pp. 1704–1716.

Jhanwar, N., S. Chaudhuri, G. Seetharaman, and B. Zavidovique "Content-based image retrieval using motif co-occurrence matrix," *Image Vision Computer*, Vol. 22, 2004, pp. 1211–1220.

Jiang, J., Y. Weng, and P. Li, "Dominant colour extraction in compressed domain", *Image and Vision Computing Journal, Elsevier*, Vol. 24, 2006, pp. 1269–1277.

Joe Stanleya, R., S. De, D. Demner-Fushman, S. Antani, and G. R. Thoma, "An image feature-based approach to automatically find images for application to clinical decision support," *Computerized Medical Imaging and Graph*, Vol. 35, 2011, pp. 365–372.

Johnson, J. D., M. H. Price, and E. K. Leiker, "Episodic retrieval involves early and sustained effects of reactivating information from encoding," *NeuroImage*, Vol. 106, 2015, pp. 300–310.

Kankanhalli, M. S., B. M. Mehtre, and H. Y. Huang, "Color and spatial feature for content-based image retrieval," *Pattern Recognition Letters*, Vol. 20, 1999, pp. 109–118.

Karakasis, E. G., A. Amanatiadis, A. Gasteratos, and S. A. Chatzichristofis, "Image moment invariants as local features for content based image retrieval using the bag-of-visual-words model," *Pattern Recognition Letters*, Vol. 55, pp. 22–27, 2015.

Kauppi, J-P., M. Kandemir, V-M. Saarinen, L. Hirvenkari, L. Parkkonen, A. Klami, R. Hari, and S. Kaskimy, "Towards brain-activity-controlled information retrieval: Decoding image relevance from MEG signals," *NeuroImage*, 2015.

Kaur, H., V. Dhir, "Local color oppugnant Mesh Extrema Patterns: A new feature descriptor for Image retrieval," *Indian Journal of Science and Technology*, 2017.

Ketz, N., R. C. O'Reilly, and T. Curran, "Classification aided analysis of oscillatory signatures in controlled retrieval," *NeuroImage*, Vol. 85, 2014, pp. 749–760.

Kherfi, M. L., D. Ziou, and A. Bernardi, "Image retrieval from the World Wide Web: Issues, techniques, and systems," *ACM Computing Surveys*, Vol. 36, 2004, pp. 35–67.

Kim, S. C. and T. J. Kang, "Texture classification and segmentation using wavelet packet frame and Gaussian mixture model," *Pattern Recognition*, Vol. 40(4), 2007, pp. 1207–1221.

Kim, N. D. and S. Udpa, "Texture classification using rotated wavelet filters," *IEEE Transactions on Systems, Man, and Cybernetics Part A: Systems and Humans*, Vol. 30, 2000, pp. 847–852.

Kinage, K. and S. G. Bhirud, "*Face Recognition using Curvelet and ICA,*" *International Conference on Image Processing, Computer Vision and Pattern Recognition, IPCV 2010*, Las Vegas, NV, 2010.

Kokare, M., P. K. Biswas, and B. N. Chatterji, "Rotation-invariant texture image retrieval using rotated complex wavelet filters," *IEEE Transactions on Systems, Man, and Cybernetics, Part B: Cybernetics*, Vol. 36(6), 2006, pp. 1273–1282. doi:10.1109/TSMCB.2006.874692

Kokare, M., P. K. Biswas, and B. N. Chatterji, "Texture image retrieval using rotated wavelet filters," *Pattern Recognition Letters*, Vol. 28, pp. 1240–1249, 2007.

Kokare, M., B. N. Chatterji, and P. K. Biswas, "A survey on current content based image retrieval methods," *IETE Journal of Research*, Vol. 48(3&4), pp. 261–271, 2002.

Kokare, M., B. N. Chatterji, and P. K. Biswas, "Cosine-modulated wavelet based texture features for content-based image retrieval," *Pattern Recognition Letters*, Vol. 25, pp. 391–398, 2004.

Koteswara Rao, L., P. Rohini, and L. Pratap Reddy, "Local color oppugnant quantized extrema patterns for image retrieval," *Multidimensional Systems and Signal Processing*, 2018. doi:10.1007/s11045-018-0609-x

Koteswara Rao, L. and D. Venkata Rao, Content based medical image retrieval using local co-occurrence patterns, ICATCCT, IEEE, 2015a.

Koteswara Rao, L. and D. Venkata Rao, "Local quantized extrema patterns for content-based natural and texture image retrieval," *Human-centric Computing and Information Sciences*, 2015b, doi:10.1186/s13673-015-0044-z

Koteswara Rao, L., D. Venkata Rao, and L. Pratap Reddy, Color based multi-directional LocalMotif XOR patterns for Image retrieval, ICATCCT, IEEE Xplore, 2015.

Koteswara Rao, L., D. Venkata Rao, and L. P. Reddy, "Local Mesh quantized extrema patterns for image retrieval," *SpringerPlus*, Vol. 5(976), 2016.

Krizhevsky, A., I. Sutskever, and G. E. Hinton, "Imagenet classification with deep convolutional neural networks." in *Advances in Neural Information Processing Systems*, 2012, pp. 1097–1105.

Kumar, J., P. Ye, and D. Doermann, "Structural similarity for document image classification and retrieval," *Pattern Recognition Letters*, Vol. 43, 2014, pp. 119–126.

Li, X., "Image retrieval based on perceptive weighted color blocks," *Pattern Recognition Letters*, Vol. 24(12), 2003, pp. 1935–1941.

Li, M. and R. C. Staunton, "Optimum Gabor filter design and local binary patterns for texture segmentation," *Journal of Pattern Recognition*, Vol. 29, 2008, pp. 664–672.

Liang, Z., Y. Zhuang, Y. Yang, and J. Xiao, "Retrieval-based cartoon gesture recognition and applications via semi-supervised heterogeneous classifiers learning," *Pattern Recognition*, Vol. 46, 2013, pp. 412–423.

Liao, S., M. W. K. Law, and A. C. S. Chung, Dominant local binary patterns for texture classification. *IEEE Transactions on Image Processing*, Vol. 18(5), 2009, pp. 1107–1118. doi:10.1109/TIP.2009.2015682

Lin, C. H., R. T. Chen, and Y. K. A. Chan, "Smart content-based image retrieval system based on color and texture feature," *Image Vision Computer*, Vol. 27, 2009, pp. 658–665.

Lin, C-H., C-C. Chen, H-L. Lee, and J-R. Liao, "Fast K-means algorithm based on a level histogram for image retrieval," *Expert Systems with Applications*, Vol. 41, 2014a, pp. 3276–3283.

Lin, C-H., H-Y. Chen, and Y-S. Wu, "Study of image retrieval and classification based on adaptive features using genetic algorithm feature selection," *Expert Systems with Applications*, Vol. 41, 2014b, pp. 6611–6621.

Lin, C-H., D-C. Huang, Y-K. Chan, K-H. Chen, and Y-J. Chang, "Fast color-spatial feature based image retrieval methods", *Expert Systems with Applications*, Vol. 38, 2011, pp. 11412–11420.

Liu, G-H., Z-Y. Li, L. Zhang, and Y. Xu, "Image retrieval based on micro-structure descriptor," *Pattern Recognition*. 2011, Vol. 44, pp. 2123–2133. doi:10.1016/j.patcog.2011.02.003.

Liu, G.-H. and J.-Y. Yang, "Image retrieval based on the texton co-occurrence matrix," *Pattern Recognition*, Vol. 41, 2008, pp. 3521–3527.

Liu, H. and C. Zhang, "Codebook design of keyblock based image retrieval," *Lecture notes in Computer Science*, Vol. 4740, 2007, pp. 470–474.

Liu, G-H., L. Zhang, Y-K. Hou, Z-Y. Li, and J-Y. Yang, "Image retrieval based on multi-texton histogram," *Pattern Recognition*, Vol. 43, 2010, pp. 2380–2389.

Liu, Y., D. Zhang, G. Lu, and W-Y. Ma, "A survey of content-based image retrieval with high-level semantics," *Elsevier Journal of Pattern Recognition*, Vol. 40, 2007, pp. 262–282.

Loupias, E., N. Sebe, S. Bres, and J.-M. Jolion, "Wavelet-based salient points for image retrieval," *Image Processing, Proceedings 2000 International Conference on*, Vol. 2, 2000, pp. 518–521. doi:10.1109/ICIP.2000.899469

172

References

Lu, Z. M. and H. Burkhardt, "Colour image retrieval based on DCT domain vector quantization index histograms," *Journal of Electronics Letters*, Vol. 41(17), 2005, pp. 29–30.

Lu, G. and S. Teng, "*A Novel Image Retrieval Technique based on Vector Quantization*," In *Proceedings of International Conference on Computational Intelligence for Modelling, Control and Automation*, 1999, pp. 36–41.

Ma, Y., X. Gu, and Y. Wang, "Histogram similarity measure using variable bin size distance," *Computer Vision and Image Understanding*, Vol. 114, 2010, pp. 981–989.

Mandal, T., W. Jonathan, and Y. Yuan, "Curvelet based face recognition via dimension reduction," *Signal Processing*, Vol. 89, 2009, pp. 2345–2353.

Mandal, T., A. Majumdar, and Q. M. J. Wu, "*Face recognition by curvelet based feature extraction*," International conference on image analysis and recognition, ICIAR, Montreal, Canada, 2007, pp. 806–817.

Mandal, M. K., S. Panchanathan, and T. Aboulnasr, "Fast wavelet Histogram Techniques for Image indexing," *Journal of Computer Vision and Image Understanding*, Vol. 75(1/2), 1999, pp. 99–110.

Manjunath, B. S. and W. Y. Ma, "Texture features for browsing and retrieval of image data," *IEEE Transactions on Pattern Analysis and Machine Intelligence*, Vol. 18, 1996, pp. 837–842.

Manjunath, B. S., J.-R. Ohm, V. V. Vasudevan, and A. Yamada, "Color and texture descriptors," *IEEE Transactions on Circuits And Systems for Video Technology*, Vol. 11, 2001, pp. 703–714.

Marcus, D. S., T. H. Wang, J. Parker, J. G. Csernansky, J. C. Morris, and R. L. Buckner, "Open access series of imaging studies (OASIS): Crosssectional MRI data in young, middle aged, nondemented, and demented older adults," *Journal of Cognitive Neuroscience*, Vol. 19(9), 2007, pp. 1498–1507.

Mehtre, B. M., M. Kankanhalli, and W. F. Lee, "Shape measures for content based image retrieval: A comparison," *Information Processing and Management*, Vol. 33, 1997, pp. 319–337.

Mehtre, B. M., M. S. Kankanhalli, and W. F. Lee, "Content-based image retrieval using a composite color-shape approach," *Information Processing and Management*, Vol. 34, 1998, pp. 109–120.

MIT Vision and Modelling Group, Vision Texture. [Online]. Available: http://vismod.media.mit.edu/pub/.

Moghaddam, H. A. and M. Saadatmand Tarzjan, *Gabor wavelet Correlogram Algorithm for Image Indexing and Retrieval*, 18th International Conference on Pattern Recognition, K.N. Toosi University of Technology, Tehran, Iran, 2006, pp. 925–928.

Moore, S. and R. Bowden, "Local binary patterns for multi-view facial expression recognition," *Computer Vision and Image Understanding*, Vol. 115(4), 2011, pp. 541–558. doi:10.1016/j.cviu.2010.12.001

Mukhopadhyay, S., J. K. Dash, and R. D. Gupta, "Content-based texture image retrieval using fuzzy class membership," *Pattern Recognition Letters*, Vol. 34, 2013, pp. 646–654.

Muller, H., W. Muller, D. McG. Squire, S. Marchand-Maillet, and T. Pun, "Performance evaluation in content-based image retrieval: Overview and proposals," *Pattern Recognition Letters*, Vol. 22(5), 2001, pp. 593–601.

Muller, H., A. Rosset, J-P. Vallee, F. Terrier, and A. Geissbuhler, "A reference data set for the evaluation of medical image retrieval systems," *Computerized Medical Imaging and Graphics*, Vol. 28, 2004, pp. 295–305.

Murala, S. and Q. M. Jonathan Wu, "Local ternary co-occurrence patterns: A new feature descriptor for MRI and CT image retrieval," *Neurocomputing*, Vol. 119(7), 2013, pp. 399–412.

Murala, S. and Q. M. Jonathan Wu, "Local mesh patterns versus local binary patterns: Biomedical image indexing and retrieval," *IEEE Journal of Biomedical and Health Informatics*, Vol. 18(3), 2014, pp. 929–938.

Murala, S., R. P. Maheshwari, and R. Balasubramanian, "Local maximum edge binary patterns: A new descriptor for image retrieval and object tracking," *Signal Processing*, Vol. 92, 2012a, pp. 1467–1479.

Murala, S., R. P. Maheshwari, and R. Balasubramanian, "Local tetra patterns: A new feature descriptor for content based image retrieval," *IEEE Transactions on Image Processing*, Vol. 21(5), 2012b, pp. 2874–2886.

Murala, S., R. P. Maheshwari, and R. Balasubramanian, "Directional binary wavelet patterns for biomedical image indexing and retrieval," *Journal of Medical Systems*, Vol. 36(5), 2012c, pp. 2865–2879.

Murala, S., R. P. Maheshwari, R. Balasubramanian, "Expert system design using wavelet and color vocabulary trees for image retrieval," *International Journal of Expert Systems with Applications*, Vol. 39, 2012d, pp. 5104–5114.

Nanni, L., A. Lumini, and S. Brahnam, "Local binary patterns variants as texture descriptors for medical image analysis," *Artificial Intelligence in Medicine*, Vol. 49(2), 2010, pp. 117–125. doi:10.1016/j.artmed.2010.02.006

NEMA-CT image database. Available [Online]: ftp://medical.nema.org/medical/Dicom/Multiframe/

Ning, J., L. Zhang, D. Zhang, and C. Wu, "Robust object tracking using joint color-texture histogram," *International Journal of Pattern Recognition and Artificial Intelligence*, Vol. 23(7), 2009, pp. 1245–1263. doi:10.1142/S0218001409007624

Ojala, T., M. Pietikainen, and D. Harwood, "A comparative study of texture measures with classification based on feature distributions," *Pattern Recognition*, Vol. 29(1), 1996, pp. 51–59.

Ojala, T., M. Pietikainen, and T. Maenpaa, "Multiresolution gray-scale and rotation invariant texture classification with local binary patterns," *IEEE Transactions on Pattern Analysis and Machine Intelligence*, Vol. 24(7), 2002, pp. 971–987.

Ojala, T., K. Valkealahti, M. Pietikainen, and M. Kokare, "Texture discrimination with multi-dimensional distributions of signed gray-level differences," *Pattern Recognition*, Vol. 34(3), 2001, pp. 727–739.

Pan, H., P. Li, Q. Li, Q. Han, X. Feng, and L. Gao, "Brain CT image similarity retrieval method based on uncertain location graph," *IEEE Journal Of Biomedical And Health Informatics*, Vol. 18(2), 2014, 574–584.

Papadopoulos, G. T., K. C. Apostolakis, and P. Daras, "Gaze-based relevance feedback for realizing region-based image retrieval," *IEEE Transactions on Multimedia*, Vol. 16(2), 2014, pp. 440–454.

Park, D. K., Y. S. Jeon, C. S. Won, and S-J. Park, "Efficient use of local edge histogram descriptor," *Proceedings of ACM Multimedia Workshops*, 2000, pp. 51–54.

Park, U., J. Park, and A. K. Jain, "Robust keypoint detection using higher-order scale space derivatives: Application to image retrieval," *IEEE Signal Processing Letters*, Vol. 21(8), 2014, pp. 962–965.

Pass, G., R. Zabih, and J. Miller, "*Comparing images using color coherence vectors*," in *Proceedings of the 4th ACM Multimedia Conference*, Boston, Massachusetts, 1997, pp. 65–73.

Peng, S., D. Kim, S. Lee, and M. Lim, "Texture feature extraction on uniformity estimation for local brightness and structure in chest CT images," *Computers in Biology and Medicine*, Vol. 40, 2010, pp. 931–942.

Pentland, A. "Fractal-based description of natural scenes," *IEEE Transactions on Pattern Analysis and Machine Intelligence*, Vol. 6(6), 1984, pp. 661–674.

Pietikainen, M., T. Ojala, T. Scruggs, K. W. Bowyer, C. Jin, K. Hoffman, J. Marques, M. Jacsik, and W. Worek, "Overview of the face recognition using feature distributions," *Journal of Pattern Recognition*, Vol. 33(1), 2000, pp. 43–52.

Pour, H. N. and E. Kabir, "Image retrieval using histograms of uni-color and bi-color blocks and directional changes in intensity gradient," *Pattern Recognition Letters*, Vol. 23, 2004, pp. 1547–1557.

Pourghassem, H. and H. Ghassemian, "Content-based medical image classification using a new hierarchical merging scheme," *Computerized Medicine Imaging and Graph*, Vol. 32, 2008, pp. 651–661.

Prasad, B. G., K. K. Biswas, and S. K. Gupta, "Region-based image retrieval using integrated color, shape, and location index," *Computer Vision and Image Understanding*, Vol. 94, 2004, pp. 193–233.

Qi, X. and Y. Han, "A novel fusion approach to content-based image retrieval," *Pattern Recognition*, Vol. 38, 2005, pp. 2449–2465.

Qiu, G., "Color Image Indexing Using BTC," *IEEE Transactions on Image Processing*, Vol. 12(1), 2003, pp. 93–101.

Raghu, P. P. and B. Yegnanarayana, "Segmentation of Gabor filtered textures using deterministic relaxation," *IEEE Transactions on Image Processing*, Vol. 5(12), 1996, pp. 1625–1636.

Rahimi, M. and E. Moghaddam, "A content based image retrieval system based on Color ton distribution descriptors," *Signal, Image and Video Processing*, Vol. 9, 2015, p. 691. doi:10.1007/s11760-013-0506-6

Rahman, M. M. and P. Bhattacharya, "An integrated and interactive decision support system for automated melanoma recognition of dermoscopic images," *Computerized Medical Imaging and Graph*, Vol. 34, 2010, pp. 479–486.

Rajavel, P., "Directional Hartley transform and content based image retrieval," *Signal Processing*, Vol. 90, 2010, pp. 1267–1278.

Rallabandi, V. R. and V. P. S. Rallabandi, "Rotation-invariant texture retrieval using wavelet-based hidden Markov trees," *Signal Processing*, Vol. 88, 2008, pp. 2593–2598.

Rao, L. K. and P. Rohini, Multiple color channel local extrema patterns for image retrieval, ICECE 2018.

Rao, A., R. K. Srihari, and Z. Zhang, "Spatial color histograms for content-based image retrieval," in *Proceedings of 11th IEEE Conference*, 1999, pp. 183–186.

Rao, K. and D. Venkata Rao, "Combination of color and Local Patterns as a feature vector for CBIR," *International Journal of Computer Applications*, Vol. 99, pp. 1–5, 2014a.

Rao, K. and D. Venkata Rao, "A feature vector for CBIR based on DLEP and Gabor features," *IJETT*, Vol. 12, 2014b.

Rao, K. and D. Venkata Rao, "Local quantized extrema patterns, Human centric Computing and information sciences," *Sciences (Springer)*, Vol. 5(26), 2015, pp. 1–24.

Rao, K., D. Venkata Rao, and P. Rohini, "*Improved DLEP as a feature vector for CBIR*," in *ICSC IEEE Conference*, 2013.

Rao, L. K., D. Venkata Rao, and P. Rohini, Integration of color and MDLEP as a feature vector in image indexing and retrieval, ICICT 2015, AISC (Springer).

Rao, K., D. Venkata Rao, and P. Rohini, *Combination of CDLEP and Gabor transform for CBIR*, AISC, Springer 2016.

Reddy, P. V. N. and K. S. Prasad, "Line Edge Binary Patterns for content-based image retrieval", *International Journal of Signal and Imaging Systems Engineering*, Vol. 5(3), pp. 227–235, 2012.

Reddy, P. V. B. and A. R. M. Reddy, "Content based image indexing and retrieval using directional local extrema and magnitude patterns," *International Journal of Electronics and Communications (AEU)*, Vol. 68(7), 2014, pp. 637–643.

Rivaz, P. D. and N. Kingsbury, "*Complex wavelet for fast texture image retrieval*," *Proceedings of IEEE International Conference on Image Processing (ICIP1999)*, Kobe, Japan, 1999, pp. 109–113.

Rui, Y., T. S. Huang, and S. F. Chang, "Image retrieval: Current techniques, promising directions & open issues," *Journal of Visual Communications & Image Representation*, Vol. 10(4), 1999, pp. 39–62.

Saadatmand, T. M. and H. A. Moghaddam, *"Enhanced Wavelet Correlogram Methods for Image Indexing and Retrieval,"* in *Proceedings of the IEEE International Conference on Image Processing*, K. N. Toosi University of Technolology, Tehran, Iran, 2005, pp. 541–544.

Sánchez, J., F. Perronnin, T. Mensink, and J. Verbeek, "Image classification with the fisher vector: Theory and practice," *International Journal of Computer Vision*, Vol. 105(3), 2013, pp. 222–245.

Santos, W. P., F. M. Assis, R. E. Souza, P. B. Santos Filho, and F. B. Lima Neto, "Dialectical multispectral classification of diffusion-weighted magnetic resonance images as an alternative to apparent diffusion coefficients maps to perform anatomical analysis," *Computerized Medical Imaging and Graph*, Vol. 33, 2009, pp. 442–460.

Sastry, C. S., M. Ravindranath, A. K. Pujari, and B. L. Deekshatulu, "A modified Gabor function for content based image retrieval," *Pattern Recognition Letters*, Vol. 28, 2007, pp. 293–300.

Shinde, S. and V. M. Gadre, "An Uncertainty Principle for Real Signals in the Fractional Fourier Transform Domain," *IEEE Transactions On Signal Processing*, Vol. 49, 2001, pp. 2545–2548.

Simou, N., T. Athanasiadis, G. Stoilos, and S. Kollias, "Image indexing and retrieval using expressive fuzzy description logics," *Signal, Image and Video Processing*, Vol. 2, 2008, p. 321. doi:10.1007/s11760-008-0084-1

Smeulders, A. W. M., M. Worring, S. Santini, A. Gupta, and R. Jain, "Content-Based Image Retrieval at the end of the Early Years," *IEEE Transactions on Pattern Analysis and Machine Intelligence*, Vol. 22(12), 2000, pp. 1349–1380.

Smith, J. R. and S. F. Chang, *"Automated binary texture feature sets for image retrieval,"* Proceedings of IEEE International Conference Acoustics, *Speech and Signal Processing*, Columbia University, New York, 1996, pp. 2239–2242.

Sorensen, L., S. B. Shaker, and M. de Bruijne, "Quantitative analysis of pulmonary emphysema using local binary patterns," *IEEE Transactions on Medical Imaging*, Vol. 29(2), 2010, pp. 559–569.

Sridhar, V., *Region-based Image Retrieval Using Multiple Features*, Department of Computing Science, University of Alberta Edmonton, Alberta, Canada, 2002.

Srinivasan, G. N. and G. Shobha, Proceedings of world academy of science, engineering and technology Vol. 36, 2008.

Starck, J. L., E. J. Candes, and D. L. Donoho, "The curvelet transform for image denosing," *IEEE Transactions on Image Processing*, Vol. 11, pp. 131–141, 2002.

Stricker, M. and M. Oreng, *"Similarity of Color Images,"* in *Proceedings of the SPIE, Storage and Retrieval for Image and Video Databases*, 1995, pp. 381–392.

Su, M-C. and C-H. Chou, "A modified version of the K-means algorithm with a distance based on cluster symmetry," *IEEE Transaction on Pattern Analysis and Machine Intelligence*, Vol. 23(6), 2001, pp. 674–680.

Su, Z., H. Zhang, S. Li, and S. Ma, "Relevance feedback in content-based image retrieval: Bayesian framework, feature subspaces, and progressive learning," *IEEE Transactions on Image Processing*, Vol. 12(8), 2003, pp. 924–937.

Subrahmanyam, M., R. P. Maheshwari, and R. Balasubramanian, "A Correlogram algorithm for image indexing and retrieval using wavelet and rotated wavelet filters", *International Journal of Signal and Imaging Systems Engineering*, Vol. 4(1), pp. 27–34, 2011.

Subrahmanyam, M., R. P. Maheswari, and R. Balasubramanian, "Directional local extrema pattern: A new descriptor for content based image retrieval," *International Journal of Multimedia Information Retrieval*, Vol. 1(3), 2012, pp. 191–203.

Sumana, I. J., M. M. Islam, D. Zhang, and G. Lu, "Content based image retrieval using curvelet transform," *Multimedia Signal Processing*, 2008, pp. 8–10.

Swain, M. J. and D. H. Ballard, "Indexing via color histograms," *Third International Conference on Computer Vision*, 1990, pp. 390–393.

Swain, M. J. and D. H. Ballard, "Color Indexing," *International Journal of Computer Vision*, Vol. 7(1), 1991, pp. 11–32.

Takala, V., T. Ahonen, and M. Pietikainen, "Block-based methods for image retrieval using local binary patterns, SCIA 2005," *LNCS* 3450, 2005, pp. 882–891.

Tamura, H., S. Mori, and T. Yamawaki, "Textural Features Corresponding to Visual Perception," *IEEE Transactions on Systems, Man, and Cybernetics*, SMC-8, 1978, pp. 460–473.

Tan, X. and B. Triggs, "Enhanced local texture feature sets for face recognition under difficult lighting conditions," *IEEE Transactions on Image Processing*, Vol. 19(6), 2010, pp. 1635–1650.

Thangaraj, M. and G. Sujatha, "An architectural design for effective information retrieval in semantic web," *Expert Systems with Applications*, Vol. 41(18), pp. 8225–8233, 2014.

Tsikrika, T., A. Popescu, and J. Kludas, "*Overview of the Wikipedia Image Retrieval task at ImageCLEF 2011*," in *the Working Notes for the CLEF 2011 Labs and Workshop*, 19–22 September, Amsterdam, The Netherlands, 2011.

Tuceryan, M. and A. K. Jain, "Texture Analysis," in *Handbook Pattern Recognition and Computer Vision*, C. H. Chen, L. F. Pau, and P. S. P. Wang, eds., Singapore: World Scientific, 1993, pp. 235–276.

Tuceryan, M. and A. K. Jain, "Texture analysis," in *The Handbook of Pattern Recognition and Computer Vision*, C. H. Chen, L. F. Pau, and P. S. P. Wang, eds. World Scientific Publishing Co., 1998, pp. 207–248.

Tuceryan, M. and A. K. Jain Chatterji, "Texture segmentation using Voronoi polygons," *IEEE Transactions on Pattern Analysis and Machine Intelligence*, Vol. 12(2), 1990, pp. 211–216.

Unay, D., A. Ekin, and R. S. Jasinschi, "Local structure-based region-of-interest retrieval in Brain MR images," *IEEE Transactions on Information Technology in Biomedicine*, Vol. 14(4), 2010, pp. 897–903.

Ursani, A. A., K. Kpalma, and J. Ronsin "*Texture features based on Fourier transform and Gabor filters: An empirical comparison*," in *Proceedings of International Conference on Machine Vision*, Islamabad, Pakistan, 2007, pp. 67–72.

Vadivel, A., S. Shamik, and A. K. Majumdar, "An integrated color and intensity co-occurrence matrix," *Pattern Recognition Letters*, Vol. 28, 2007, pp. 974–983.

Veltkamp, R. and M. Tanase, Content-based image retrieval systems: A survey, Technical Report UU-CS-2000–34, Department of Computing Science, Utrecht University, 2002.

Verma, M. and B. Raman, "Local tri-directional patterns: A new texture feature descriptor for image retrieval," *Digital Signal Processing*, Vol. 51, 2016, pp. 62–72.

Verma, M. and B. Raman, "Local neighborhood difference pattern: A new feature descriptor for natural and texture image retrieval," *Multimedia Tools and Applications*. 2017. doi:10.1007/s11042-017-4834-3

VIA/I-ELCAP CT Lung Image Dataset. Available [Online]: http://www.via.cornell.edu/databases-/lungdb.html.

Vipparthi, S. K., S. Murala, A. B. Gonde, and Q. M. Jonathan Wu, "Local directional mask maximum edge patterns for image retrieval and face recognition," *IET Computer Vision*, Vol. 10(3), 2016, pp. 182–192.

Vipparthi, S. K., S. Murala, S. K. Nagar, and A. B. Gonde, "Local gabor maximum edge position octal patterns for image indexing and retrieval," *Neurocomputing*, Vol. 167, 2015, pp. 336–345.

Vipparthi, S. K. and S. K. Nagar, "Expert image retrieval system using directional local motif XoR patterns," *Expert Systems with Applications*, Vol. 41, 2014a, pp. 8016–8026.

Vipparthi, S. K. and S. K. Nagar, "Multi-joint histogram based modelling for image indexing and retrieval," *Computers and Electrical Engineering*, Vol. 40, 2014b, pp. 163–173.

Vipparthi, S. K. and S. K. Nagar, "Integration of color and local derivative pattern features for content-based image indexing and retrieval," *Journal of the Institution of Engineers (India): Series B*, 2015, pp. 1–13.

Vipparthi, S. K. and S. K. Nagar, "Local extreme complete trio pattern for multimedia image retrieval system," *International Journal of Automation and Computing*, Vol. 13, 2016, p. 457. doi:10.1007/s11633-016-0978-2

Vo, A. P. N., T. T. Nguyen, and S. Oraintara, "Texture image retrieval using complex directional" in *Proceedings of the ISCAS'06*, 2006, pp. 5495–5498.

Walia, E., A. Goyal, and Y. S. Brar, "Zernike moments and LDP-Weighted patches for content based image retrieval," *SIVP*, Vol. 8, 2014, p. 577. doi:10.1007/s11760-013-0561-z

Wang, H., M. Ullah, A. Klaser, I. Laptev, and C. Schmid. Evaluation of local spatio-temporal features for action recognition, BMVC, 2009.

Weszka, J. S., C. R. Dyer, and A. Rosenfeld, "A comparative study of texture measures for terrain classification," *IEEE Transactions on Systems, Man and Cybernetics*, Vol. 6(4), 1976, pp. 269–285.

Wu, W-J., S-W. Lin, and W. K. Moon, "Combining support vector machine with genetic algorithm to classify ultrasound breast tumor images," *Computerized Medical Imaging and Graph*, 2012. doi:10.1016/j.comp-medimag.2012.07.004.

Yang, X., X. Qian, and T. Mei, "Learning salient visual word for scalable mobile image retrieval," *Pattern Recognition*, Vol. 48(10), pp. 3093–3101, 2015.

Yang, Y., L. Yang, G. Wu, and S. Li, "Image relevance prediction using query-context bag-of-object retrieval model," *IEEE Transactions on Multimedia*, Vol. 16(6), 2014, pp. 1700–1712.

Yao, C-H. and S-Y. Chen, "Retrieval of translated, rotated and scaled color textures," *Pattern Recognition*, Vol. 36, 2003, pp. 913–929.

Yoo, H. W., H. S. Park, and D. S. Jang, "Expert system for color image retrieval," *Expert Systems with Applications*, Vol. 28, 2005, pp. 347–357.

Yu, S.-N., C-T. Chianga, and C-C. Hsieh, "A three-object model for the similarity searches of chest CT images," *Computerized Medical Imaging and Graph*, Vol. 29, 2005, pp. 617–630.

Zhang, B., Y. Gao, S. Zhao, and J. Liu, "Local derivative pattern versus local binary pattern: Face recognition with higher-order local pattern descriptor," *IEEE Transactions on Image Processing*, Vol. 19(2), 2010a, pp. 533–544.

Zhang, Q. and E. Izquierdo, "Histology image retrieval in optimized multifeature spaces," *IEEE Journal of Biomedical and Health Informatics*, Vol. 17(1), 2013, pp. 240–249.

Zhang, X., W. Liu, M. Dundar, S. Badve, and S. Zhang, "Towards large-scale histopathological image analysis: Hashing-based image retrieval," *IEEE Transactions on Medical Imaging*, Vol. 34(2), 2015, pp. 496–509.

Zhang, D. and G. Lu, "Shape-based image retrieval using generic Fourier descriptor," *Signal Processing Image Communication*, Vol. 17, 2002, pp. 825–848.

Zhang, D. and G. Lu, "A comparative study of curvature scale space and Fourier descriptors for shape-based image retrieval," *Journal of Visual Communication and Image Representation*, Vol. 14, 2003, pp. 41–60.

Zhang, B., L. Zhang, D. Zhang, and L. Shen, "Directional binary code with application to Poly U near-infrared face database," *Pattern Recognition Letters*, Vol. 31, 2010b, pp. 2337–2344.

Zhao, G. and M. Pietikainen, "Dynamic texture recognition using local binary patterns with an application to facial expressions," *IEEE Transactions on Pattern Analysis and Machine Intelligence*, Vol. 29(6), 2007, pp. 915–928.

Zhou, N. and J. Fan, "Automatic image–text alignment for large-scale web image indexing and retrieval," *Pattern Recognition*, Vol. 48, 2015, pp. 205–219.

Zhu, L., A. Zhang, A. Rao, and R. Srihari, *"Keyblock: An approach for content-based image retrieval,"* in *Proceedings of ACM Multimedia*, Los Angeles, CA, 2000, pp. 157–166.

Zuo, J. and D. Cui, *"Retrieval oriented robust image hashing,"* in *Proceeding International Conference on Industrial Mechatronics and Automation*, Chengdu, 2009, pp. 379–381.

# Index

Printed in the United States
by Baker & Taylor Publisher Services